Praise for *Showdow.* W9-BOO-094

"Best account yet of how Obama strategizes."

—Jefferson Morley, Salon.com

"A dramatic behind-the-scenes account of the decision making that occurred within the Obama White House following the disastrous-for-Democrats 2010 midterm election. . . . Shows how Obama got his groove back in time for the 2012 campaign."

—Huffington Post

"Through the lens of policy battles, you get a remarkably vivid pen portrait of Obama himself and how he thinks."

—Kevin Drum, *Mother Jones*

"Fine book about how the Obama administration maneuvered to its advantage in the face of the catastrophic losses of the 2010 midterm elections."

—Charles P. Pierce, *Esquire*

"Why did Obama and Democrats pivot so hard towards deficit reductions and away from job creation after the 2010 elections? We now have a book that sheds some fresh light on what drove this pivot. . . . Corn reports on a number of behind the scene discussions. . . . Illuminating."

—Greg Sargent, *Washington Post*

"Veteran Washington journalist Corn highlights the Obama presidency's most important and dramatic events. . . . The book brings a different perspective just as Obama fights for re-election."

—*Publishers Weekly*

ALSO BY DAVID CORN

*The Lies of George W. Bush: Mastering the
Politics of Deception*
Blond Ghost: Ted Shackley and the CIA's Crusades
Deep Background: A Novel

BY MICHAEL ISIKOFF AND DAVID CORN

*Hubris: The Inside Story of Spin, Scandal,
and the Selling of the Iraq War*

SHOWDOWN

———✦———

*The Inside Story of How Obama
Battled the GOP to Set Up
the 2012 Election*

DAVID CORN

wm
WILLIAM MORROW
An Imprint of HarperCollins*Publishers*

All insert photographs were taken by Pete Souza/White House Photos, except the photograph on the bottom of page 2, which was taken by Chuck Kennedy/White House Photos.

SHOWDOWN. Copyright © 2012 by David Corn. All rights reserved. Printed in the United States of America. No part of this book may be used or reproduced in any manner whatsoever without written permission except in the case of brief quotations embodied in critical articles and reviews. For information address HarperCollins Publishers, 10 East 53rd Street, New York, NY 10022.

HarperCollins books may be purchased for educational, business, or sales promotional use. For information please write: Special Markets Department, HarperCollins Publishers, 10 East 53rd Street, New York, NY 10022.

A hardcover edition of this book was published in 2012 by William Morrow, an imprint of HarperCollins Publishers.

FIRST WILLIAM MORROW PAPERBACK EDITION PUBLISHED 2012.

Library of Congress Cataloging-in-Publication Data has been applied for.

ISBN 978-0-06-210800-5

12 13 14 15 16   OV/RRD   10 9 8 7 6 5 4 3 2

For Welmoed, Maaike, and Amarins

It is silly not to hope.
—Ernest Hemingway, *The Old Man and the Sea*

CONTENTS

SHOWDOWN

Introduction

BENDING THE ARC

IT WAS THE FIRST SATURDAY MORNING OF DECEMBER 2010. President Barack Obama was in the Oval Office with Vice President Joe Biden, surrounded by his top aides: message guru David Axelrod, press secretary Robert Gibbs, acting chief of staff Pete Rouse, longtime friend and adviser Valerie Jarrett, Treasury secretary Timothy Geithner, National Economic Council chief Larry Summers, budget director Jack Lew, economic adviser Gene Sperling, and others. Once again, the nation's chief executive had a tough presidency-defining decision to render.

A month earlier, Obama had experienced what he had accurately dubbed "a shellacking" at the polls: in the first midterm elections of his presidency, the Republicans had won sixty-three seats in the House of Representatives, seizing control of that body. In the Senate, the GOP had also advanced, cutting the Democrats' edge by two-thirds. With unemployment still terribly high—the ongoing result of the financial collapse that had occurred during George W. Bush's administration—the president and his Democratic comrades had been soundly repudiated by the voters, with an electoral loss greater than many observers had predicted and far worse than Obama and his team had anticipated.

Whatever ideas Obama had for bolstering the still-flagging economy, for funding his preferred public investments for the future, for responding to climate change and other challenges, the House Republicans—dominated by the extremist, Obama-hating Tea Party wing of their party—would be there to thwart him.

Yet the president was determined to demonstrate that he was still a leader who could deliver results and, yes, hope. And this weekend morning, he and his aides had gathered in his office to consider a private—and politically risky—proposal the vice president was bearing.

FOR MONTHS, A BATTLE HAD BEEN BREWING OVER A FUNDAMENtal political and policy question: Should wealthy Americans, at this time of economic peril, continue to receive the generous tax-cut bonus Bush had awarded them in his first term? The Bush tax cuts were due to end on December 31, 2010. If Obama and Congress did nothing, taxes would not only go up on the wealthy—but also on middle-class Americans covered by the Bush cuts. That would cause personal hardship for millions of Americans. Worse, Obama and his economic team worried, this would likely damage the tenuous economic recovery that was sputtering along.

Obama had been unambiguous: he favored continuing the lower rates for middle-class earners, and allowing the top-income rates to return to the pre-Bush levels of the 1990s (when the economy was zipping along rather well). Smothering the Bush tax cuts for the well-to-do had been a prominent promise—and a major applause line—during his 2008 run for the presidency.

But the filibuster-wielding Republicans were not interested in decoupling the middle-class cuts from those for the rich. They wanted to see all the Bush tax cuts extended. They were willing to hold the middle class hostage: no cuts for these Americans, unless the extra breaks for those making over $250,000 a year continued—even if the tax cuts for the rich would add an estimated $700 billion to the deficit over the coming decade.

Obama knew many of his Democratic allies on Capitol Hill and progressive activists and commentators were spoiling for a fight—particularly in the aftermath of the midterms debacle. Such a confrontation would define the difference between the parties; it could revive Democratic prospects. If it could be won. Obama and his aides had long wondered whether enough House and Senate Democrats would stand fast in this sort of face-off and resist the hackneyed GOP charge that Democrats favored tax hikes. Not likely, Obama and his crew had concluded. They assumed a standoff would probably be lost due to Democratic weakness. ("There's one thing I know," Biden told the president more than once, "and that's the Senate, and I've never seen the Democrats hold all together on a tough tax vote.")

Another factor was shaping Obama's thinking: he was the president of all Americans—even those who had not voted for him—not merely the leader of a political party. He had to take into account the immediate well-being of millions of citizens. He was playing dice with their money—and he was reluctant to gamble on a fight that could end up with no agreement and, consequently, higher tax bills for middle-class workers. The imperatives of governing, he knew too well, don't always line up with those of politics.

And this dilemma was nothing new. "In the first two years of the administration," a former senior Obama White House official later said, "there was a split between those who believed in fighting for things even if you know you're going to lose and those who said get the best thing possible passed. As for the president, he really is a stone-cold progressive. But he believes you have to get stuff done."

The president was aware of the rap on him. He realized that if he cut a deal, he would be accused of capitulating to political terrorists, of being too enamored with consensus building, of rewarding the bad behavior of the Republicans who had been trying to derail his presidency from the first moments of his administration. He would be called a sellout.

Polls showed that most Americans supported ending the Bush tax-cut bonus for the most fortunate. But Obama was operating within a conservative and hostile political environment that would soon be more conservative and more hostile to him and his policies. His approval ratings were not strong, and the recent elections had created a sense of political momentum for the opposition.

In previous meetings, Obama had told his aides: if you can outline a path for me in which a bare-knuckles political fight yields a better outcome for the Americans we most care about, I'll choose that course. No one spoke up. They all realized that in a few weeks' time, the Republicans would be running the House. Winning a legislative battle over taxes—or anything else—at that point would be nearly impossible for the White House.

The president was at the intersection of crosscutting currents. He had an obligation to govern and do right by most Americans. He had a campaign promise to keep—one based on a principle he held dear. And he had his own political future to consider, while meeting his obligation to lead the besieged Democratic Party into what appeared to be an uphill 2012 battle.

Progressives could howl about yielding ground to the Republicans on Bush's tax cuts. But the biggest setback for liberals would be a Republican in the White House in 2013. Forging a compromise with obdurate GOPers might well prevent the economy from worsening—a necessary condition for Obama's reelection—and improve the president's standing among the crucial block of independent voters, many of whom had become skeptical of the president and his policies.

Obama remained caught between the two halves of his political base: independents looking for a nonpartisan leader who could stay above the fray and force Washington to do its business and progressives yearning for an ideological champion who would charge into that fray, dispatch his Republican foes, and triumph. Obama could not satisfy both groups at once—especially while dealing with opponents he publicly likened to hostage takers.

The president—at this juncture—was leaning toward cutting a deal. But he had an idea: if the Republicans were holding middle-class tax cuts hostage, he would hijack the entire tax-cut fight and exploit it to advance progressive policies that assist middle-class families and juice up the economy. While the rest of the world would be watching a scuffle over the Bush tax cuts, Obama would be maneuvering on a different field—a field, for better or worse, not widely recognized. He was aiming not for a zero-sum political victory—denying the Republicans their cherished tax breaks for the well-heeled. His goal was a victory he couldn't otherwise achieve: more federal support for financially stressed-out Americans *and* hundreds of billions of dollars in another shot of stimulus for the economy. But, he would tell aides, don't use the s-word. It was as if he had a secret.

THAT SATURDAY MORNING, BIDEN PLACED BEFORE OBAMA AN offer from Senator Mitch McConnell, the top Republican in the upper chamber, for a tax-cut deal that included preserving the high-end breaks, and Obama had to make a call: whether or not to pursue a compromise that would require him to concede on a matter of principle in order to gain on grounds of policy.

In the Oval Office, Obama was surrounded by reminders of Martin Luther King Jr. He had placed a bust of the inspiring civil rights leader on a table across the room from his desk. On the wall was a gift from a friend: a framed program from King's famous 1963 March on Washington, where he had proclaimed, "I have a dream." On the floor was a rug that included a saying King often cited: "The arc of the moral universe is long, but it bends toward justice." These were not totems of political compromise. Yet as the president and his closest advisers in this room evaluated the McConnell offer, they were contemplating yielding to Republican intransigence—albeit to achieve a higher goal.

In a 2004 interview, Obama, then running for the US Senate, had explained his days as a state senator bent on concocting strong compromises: "You can't always come up with the optimal solution, but you can usually come up with a better solution. A good compromise, a good piece of legislation is like a good sentence. Or a good piece of music. Everybody can recognize it. They say, 'Huh. It works. It makes sense.' That doesn't happen too often, of course, but it happens."

Obama was now trying to write a symphony. He was aiming for a compromise—to help those millions of Americans, to do what he could to bolster the economy, to score points with independents, and to demonstrate that he still had the power and savvy to govern and make things happen. In the year ahead, the opportunities to do all that would be severely limited. That's why this compromise had to be right.

"Let's make this work," he told his aides. His presidency could depend on it.

After the midterm elections, Obama needed to reboot. The coming year would be a continuous confrontation with the newly empowered Republicans of Capitol Hill. Obama would have to figure out how to govern within narrow and less hospitable confines, how to produce results from a divided government, and how to win the brewing political battle over fundamental notions, such as the role of government. In this period, he would be setting up the grand choice of 2012, when the nation would experience a presidential election covering a stark political divide, with government-is-the-problem Republicans fighting a cage match against Obama, who still believed government ought to be used to nudge the nation in a better direction and guide the country's difficult transition toward a vibrant twenty-first-century economy. Each side would be wrestling for the political soul of the United States.

Though at this point Obama appeared to have his back against the wall, much had gone right during his first two years. He had accomplished a great deal. In his first days, he banned the use of torture, what the Bush-Cheney crowd had called "enhanced interrogation techniques." (He also ordered the shutdown of the controversial Guantánamo prison—which would be difficult to achieve.) Weeks into office, Obama enacted the American Recovery and Reinvestment Act, the stimulus bill that shot $787 billion into the economy. The bill—though much derided by Republicans as a symbol of excessive government spending— would lift employment by up to three million jobs (by creating and saving jobs), according to the nonpartisan Congressional Budget Office.

The next year, he passed historic health care legislation that would provide medical insurance coverage to tens of millions of Americans and prohibit abusive practices of insurance companies, while compelling these businesses to spend 80 to 85 percent of their premiums on medical care, as opposed to administrative costs and executive bonuses. (Obama's top advisers had repeatedly counseled him to drop or downsize his ambitious health care effort, and Obama had refused. As Bill Burton, the former deputy press secretary, put it, "There was a point when the president was the only person in the White House who thought that health care would happen.")

Obama had enacted Wall Street reform and created the Consumer Financial Protection Bureau, which would police banks, credit card companies, mortgage lenders, and other financial firms that tried to hoodwink consumers. He presided over the second half of the TARP bailout of the financial sector and a government rescue of the US auto industry—policies that were widely criticized (justifiably so on certain particulars) but that essentially succeeded in preventing these major economic players from collapsing.

He dramatically increased the diversity of judicial nominees and appointed two women, including the first Latina, to the Supreme Court. He removed the ban on federal funding for family planning groups that promote abortion rights and work overseas. He also increased Pell grants, expanded health care coverage to benefit four million more children, provided states aid to save about 150,000 teacher jobs, reformed the student loan program, forced new consumer protections on credit card companies, granted the Food and Drug Administration the authority to regulate tobacco products, created incentives for clean energy jobs, and enacted legislation that would allow women workers to challenge instances of pay discrimination. Obama kept his vow to reduce US troops in Iraq.

This was a period of tremendous and historic productivity. Yet left-of-center critics had consistently registered complaints about the former candidate of hope: The stimulus was not nearly large enough, and Obama had not subsequently pressed for bold and extensive jobs initiatives. His appointments were too centrist. The health care reform bill was a hard-to-understand and poorly sold mishmash and lacked the so-called public option. Obama's new Wall Street rules were not tough enough, and he showed little taste for blasting Big Finance and holding it fully accountable for the economic crash caused by its rapacious practices. His programs to address the mortgage crisis—which was at the center of the nation's economic malaise—were not sufficient or fully effective.

Other progressive gripes piled up. Obama failed to push the world toward an enforceable climate change treaty at the 2009 Copenhagen summit (and instead patched together a voluntary accord among the world's leading emitters of climate change gases). He had not nominated enough progressive judges to counterbalance President Bush's stuffing of the judiciary with conservative jurists. His troop withdrawal in Iraq was slow. He dispatched more troops to Afghanistan before committing to a gradual drawdown. Gitmo wasn't closed. He continued controversial national security policies from the Bush-Cheney years (in some instances modifying them). He opposed efforts to investigate the Bush-Cheney administration.

His fellow Democrats on the Hill complained that the president and his crew, in the face of an unrelenting fusillade from the Right, had failed to sell their joint accomplishments to the public, losing the message wars over the stimulus, health care reform, the new Wall Street rules, and other initiatives.

Obama and his advisers, following the midterms debacle, readily acknowledged they had flopped as marketers. They claimed to have a good excuse: in the face of multiple crises, they were intently focused on devising policies and guiding them through Congress. While so occupied, Obama and his aides had a difficult time convincing independent voters (and others) that they had changed Washington for the better.

Obama's legislative wins looked messy and highly partisan. That was largely due to Republican wrench throwing, but Obama had not established a context in which Republicans would pay a political price for their unrelenting obstructionism.

The president had made Washington more productive, but he had not altered its ugly ways or toxic atmosphere.

"Emergencies got in the way of coherent messaging," David Axelrod said. "And we were operating in a political environment when every day is an election day and covered as such." But politics is selling—and Obama and his strategists were repeatedly bested by the Republicans in the struggle to control the political narrative of the moment.

How could Obama and his top aides—intelligent people who had done so well promoting a message in 2008—have failed so miserably at such a basic task once they hit the White House?

Asked this question, one Obama confidante said that there was no good answer.

OBAMA'S ALLIES HAD ANOTHER COMPLAINT. THE REPUBLICANS constantly tried to block Obama's programs (often relying on overheated rhetoric or falsehoods) and allied themselves with conservatives who accused Obama of being a foreign-born secret socialist Muslim hell-bent on weakening the United States of America and diminishing the wealth and freedom of its citizens—or something like that; the conspiracies varied. Yet Obama insisted on treating his Republican opposition respectfully and finding avenues for cooperation.

He rarely punched back as hard as he could, and this puzzled some supporters, who feared that say-anything Republicans were too often getting the better of Obama in the nation's endless political squabbles. They questioned his negotiating skills and his fortitude. After the midterm elections, well-known Democratic strategist James Carville, speaking at a breakfast with reporters, cracked, "If Hillary gave up one of her balls, and gave it to Obama, he'd have two."

"It frustrates people that Obama is unwilling to go for the cheap political point," Axelrod remarked. "He cares more about the larger result and that's antithetical to the gestalt of Washington. We take a lot of hits for that."

Obama the candidate had denounced the partisan bickering of Washington. He promised to remake the nation's capital so that its denizens did the people's business, rather than engage in continuous, cable TV–enabled mud wrestling. That was a heavy-duty and hard-to-fulfill promise, but it had been a large part of his appeal to moderate and independent voters (and some Republicans).

This vow also reflected a dominant part of his own nature: Obama is a consensus seeker, even when confronted with ferocious political and ideological opponents. "At his core is a search for common ground," remarked a former senior Obama administration official who worked on his 2008 campaign. "I read his book *The Audacity of Hope* and kept saying to myself—no matter what issue he was addressing—that's his answer. He is always trying to get consensus—but consensus that advances a principle he cares about."

Another former Obama White House aide put it this way: "Bill Clinton every day of his life got up thinking, 'What is the best argument I can make today to defeat the Republicans?' Obama doesn't think that way. He's not a superpartisan person. He's not in it for the fight. He doesn't wake up every day thinking how to defeat [Republican leader John] Boehner. It's, 'How can I get Boehner to agree? How can I have a reasonable conversation with Boehner? How can I make Boehner see it's in both of our interests to do this or that?'"

Obama the president had turned out to be far more pragmatic than Obama the idealistic candidate of change. "Obama is an idealist in terms of what he wants to get done," said a former aide who worked on the campaign and in the White House.

"He's a pragmatist in terms of how he gets it done." The politics of hope had become the politics of the possible.

Critics on the left and within the so-called netroots also griped that the White House had opted for conventionality, leaving millions of supporters at the side of the road. Obama had not channeled the popular engagement of the 2008 campaign into a force he could use to pressure Washington. He mostly eschewed movement-based politics and largely conducted his presidency in a traditional manner, when necessary brokering deals in private.

White House aides were sensitive about this criticism, but they had a counterargument. "You do not have the capacity to involve the public in the same way when governing as you do on a campaign, when there are opportunities to stuff envelopes, knock on doors," one senior administration official maintained.

A top Obama adviser noted, "A presidential campaign is self-contained. It's about the relationship between the candidate and the people. You don't have to get anything done. Governing requires doing things and the involvement of more people within the government. In a campaign, you can wake up one day and say let's do X, and then do it. As a president, you wake up, and you depend on agencies, Congress, and others. In governing it's much more of a challenge to bring people along with you. You get to the White House, and it's all top-down. It's not surprising that supporters were not buying this was still a people-powered operation."

Obama especially hated being hit from the Left on health care. "Nothing frustrated the president more than progressives saying health care was a piece of shit," a former top adviser said. "We got all these people covered." Obama saw the Right bash him for a supposed government overreach that threatened individual freedom, and the Left whine about the bill not going far enough. This cacophony, he fretted, was drowning out this major accomplishment.

Obama's political problems extended beyond his relationship with disenchanted liberal activists or disheartened congressional Democrats. All that could always be patched up. At this point in his presidency, Obama had to think about how to handle the emboldened obstructionists of the GOP. He would have to gain control once again of the nation's political tale. He had commanded it during the 2008 campaign. Now he would have to wrest it back from Tea Party Republicans who were sweeping into Washington waving the banner of small government and spending cuts. They had a simple story to tell: the economic woes of the nation were due to excessive spending of the government. That was inaccurate and ahistorical. But this explanation served their agenda of cut, cut, cut—and the agenda of the conservative donors, groups, and foundations that had rushed to underwrite and organize the Tea Party into an anti-Obama electoral force.

Obama had to provide counterprogramming. Calculating how to govern responsibly *and* engage in effective political warfare during a time of economic crisis—which Obama was currently doing with the tax-cut issue—would only become more challenging very soon.

AT THIS CROSSROADS, OBAMA WAS DOING A FAIR SHARE OF PONdering. He believed that much of the criticism he drew was from commentators and politicos overly fixated on the short term. He derided the constant cable television chatter and told friends and colleagues he didn't heed the instant verdicts rendered by pundits on a daily—or hourly—basis. But at times he would cite columns that had ticked him off.

Obama knew people wanted change and results quickly—especially when unemployment remained high and economic growth was slow. This was particularly true for independent voters whose loyalties, by definition, were fickle. But, he told himself, keep an eye on the horizon.

Obama wanted to guide the country into its next phase. But events had conspired against him. He would routinely note that he had been handed "a real shitty deal" when he entered the White House. He had made difficult, unpopular decisions to continue the TARP bailout of the financial firms and then to rescue the failing auto companies. Though these moves prevented the loss of millions of jobs, there was little political reward for having prevented the economic implosion from becoming worse.

The president was keenly attuned to what he considered a crucial fact: the country was sharply divided. Though his victory in 2008 had seemed decisive, he had not forgotten that it was a 53 to 46 percent win in the popular vote—and that was with the much-mocked Sarah Palin on the opposing ticket. As far as he saw it, his own triumph did not signify a basic shift in the nation's political culture. Changing the fundamental mind-set of the country regarding government, taxes, foreign policy, education, the environment—this was a thirty-year mission. Roosevelt's New Deal sensibility had defined politics for four decades, until Ronald Reagan reshaped the political culture in the 1980s and bolstered an antigovernment backlash. Obama was hoping to lead a new transition. He had concluded this could only be done gradually, in a manner that drew popular support from an American public that was often so at odds with itself politically, ideologically, and culturally.

At the same time, Obama would have to overcome obstacles set up by a Republican Party that was willing to be irresponsible. Many of its new House members would consider blocking unemployment benefits a victory, or shutting down the government an achievement to brag about. As a president who cared about what he could accomplish for low-income and middle-class families, he realized, he would be at a disadvantage in negotiations with Tea Party–driven Republicans. He was thinking about how to strike the right balance of compromise and confrontation.

Obama was indeed burdened with a rotten economy not of his making, as he regularly mentioned. But he also knew he had slipped up. In conversations with advisers, he acknowledged that the broader narrative had been lost. It was his job to tell his own story. He had done that so well during the campaign. But not as a president.

In failing to do this, he had allowed a meaner, darker vision to gain traction. His antagonists, using Obama's own accomplishments, contended that government was the problem, that the economy was troubled because of wild spending in Washington and not due to Big Finance shenanigans and malfeasance; his foes took aim at the notions of communal action and collective responses. On Obama's watch, these larger progressive themes seemed to lose ground.

Politics is not a meritocracy, and Obama realized no president was guaranteed credit or lucky bounces. But at times he couldn't help feeling, as he told one associate, a kinship with the protagonist in Ernest Hemingway's *The Old Man and the Sea*. He had, against tremendous odds, caught a big fish, but on the long voyage back to shore, his prized catch had been picked to pieces by sharks.

Moreover, the president was aware that sometimes it looked— and it certainly felt—as if he and his aides were just grinding it out. He had concluded that he needed to inject some poetry into the coming year, for people don't live on prose alone. He assumed that the next two years would not be about legislating, but about storytelling. About the nature of America—and what's best for the country in the long run. This was what he expected to be at the heart of his reelection campaign.

He figured he possessed a fighting chance—in part because the Republicans, though ascendant at the moment, were in a jam of their own devising. They were calling for reducing deficits and the national debt, yet they were eager to hand tax breaks to the wealthy. That central contradiction made them vulnerable. Plus, the Tea Partiers in their ranks would be calling for extreme actions, such as the elimination of Medicare or Social Security, and driving the party to the far right. The Republicans would overreach. They would offer Obama and his party room to recover. But the president and his strategists realized that many voters would not be willing to listen to anything he had to say if they were not experiencing an improving economic reality. Without better economic conditions, Obama's next round of hope and change would be a tough sell.

The tax-cut deal would be the start to an arduous year. It might also be Obama's last chance to enact major policies that could have a positive impact on the frail economy. In the previous two years, Obama had often won legislative battles while losing the corresponding political fights. In the coming year, there likely would be no major legislative successes, and he'd have to flip the script and become better—much better—at defining the public debate, and winning it.

Obama viewed this lame-duck session and the subsequent stretch as an opportunity to shape the terrain for the colossal clash of 2012. He had a basic plan: resolve the unavoidable budgetary issues with the Republicans, clear those matters off the table, and then compete with the GOP over fundamental messages.

It was his hope that the year ahead would be a time of debate over opposing values. He could not know what obstacles would arise. But every twist and turn—and there would be many— would hold much at stake for him, and anyone who yearned for the Obama presidency to succeed.

Chapter One

BEFORE THE SHELLACKING

"We knew this wouldn't be easy."

It was mid-May 2010—half a year before his big decision on a tax-cut deal—and President Barack Obama was addressing hundreds of high-dollar Democratic contributors (celebrities, financiers, power players) from a stage in a ballroom at the ritzy St. Regis hotel in New York City. The goal, as at every fundraiser, was to squeeze them for money. He had to show them something special, reveal a little bit of himself and his thoughts, so they would feel part of his presidency and attuned to his vision.

Obama was not an enthusiastic fund-raiser. Some politicians actually enjoy grubbing for money—not him. After he became president, some big-dollar donors grumbled about the lack of attention the White House paid to its wealthy contributors. There weren't sufficient invitations to the White House or requests for advice. Fair or not, this was regarded as a reflection of Obama's own jaundiced view toward the money chase. But when Obama wanted to reach out and touch the donors, he could. And now he truly needed to mobilize—and motivate—his party's money people.

Two months earlier, the Democrats in Congress had finally passed Obama's comprehensive but controversial health care overhaul. Tens of millions of Americans would eventually obtain health insurance coverage (which would be subsidized for low-income recipients); insurance companies would be prevented from denying coverage to customers with preexisting conditions; young adults could remain on their parents' plans; older Americans would receive additional assistance in paying for medicine.

But the passage of this landmark law, which included an individual mandate that would compel people to obtain coverage (to the horror of conservatives), had not won Obama much, if any, political credit. The whole effort had exhausted Washington and alienated independent voters, especially those who wondered if Obama's devotion to this reform had come at the expense of repairing the still-tattered economy.

The president had been left with a let-me-explain, hard-to-understand win that had prompted rage on the right and fired up the Tea Party forces now threatening his party.

Unemployment had topped 9.5 percent for nearly a year, and the president's approval rating was just below 50 percent. Congressional Democrats were faring worse. In Gallup polls, approval of the Democratic Party, which controlled both the Senate and the House, had plummeted in the past year from 55 to 41 percent. Republicans, who had embraced a consistent stance of anti-Obamaism, had crept up from 34 percent to 42. One survey showed that more than 70 percent of the public disapproved of the Congress.

All this confirmed the obvious: Obama and the Democrats were hurtling toward a cliff in the 2010 midterm elections. And for months leading up to the St. Regis fund-raiser, strategists at the Democratic Congressional Campaign Committee, the outfit tasked with preserving the Democratic majority in the House, had been urging the White House to unleash the president.

They wanted Obama to mount up and lead the charge against the say-no-to-everything Republicans. Obama's aides, though, were not eager to set Obama loose as a partisan bulldozer—which aggravated the Democratic strategists. "They'd say, 'Look, the president won the 2008 election because he was beyond politics,'" one House Democrat recalled.

But at this swanky $15,000-a-person event featuring food prepared by chefs of the French Culinary Institute, with actors Matthew Broderick and Sarah Jessica Parker in the crowd, Obama was delivering his first campaign speech of the season. After those months of back-and-forth between the White House and the DCCC, he was finally leaping into the 2010 mix. With gusto.

"I hope you knew" that it would be tough, he told the faithful. "Because I told you. If you didn't know, you weren't paying attention at my inauguration address. Remember that? Washington, couple million people, really cold?"

He pounced on the Republicans: "You would have thought at a time of historic crisis that Republican leaders would have been more willing to help us find a way out of this mess—particularly since they created the mess."

Unsheathing the biting sense of humor he only rarely displayed in public, Obama went on: "We got our mops and our brooms out, we're cleaning stuff out, and they're sitting there saying, 'Hold the broom better.' 'That's not how you mop.' Don't tell me how to mop. Pick up a mop!" The well-heeled Democratic loyalists laughed loudly.

Obama insisted that the GOP strategy had been "to gum up the works; to make things look broken." And "if it didn't work out so well, maybe the other side would take the blame." He was on a roll: "So after they drove the car into the ditch . . . now they want the keys back. No! You can't drive! We don't want to have to go back into the ditch! We just got the car out!"

Representative Chris Van Hollen, the bright-eyed congress-man from Maryland who headed the DCCC, flashed a thumbs-up at House Speaker Nancy Pelosi, who was also at the dinner. The president was using the car-in-the-ditch line Van Hollen had been quipping for months. But Van Hollen wasn't thrilled merely because Obama had cribbed a talking point from him. More important, the president was at last doing what House Democratic strategists had yearned for: pounding the Republicans.

But was this coming too late? House Democratic strategists would be disappointed that the president did not constantly pummel the Republicans over the coming weeks. This going-back-into-a-ditch point, they believed, could not be repeated often enough. Yet Obama tended to push this theme mostly at fund-raising events.

At the White House, Obama's aides believed the president was indeed accommodating congressional Democratic desires.

"The back-versus-forward message was hatched around a table with Nancy Pelosi, her leadership team, and Axe," recalled a senior White House official. "It was not the message we would have chosen. We would have preferred a message closer to our 2008 message of reforming Washington. But this is what the House and Senate felt made sense, and we wanted to do what they wanted."

Even so, the president's speech at the St. Regis included little on what he and the Democrats would do in the months ahead to improve the economic fortunes of the millions of Americans still facing or fearing rough circumstances.

"At that time in general," a senior House Democrat later commented, "there was a question whether Obama was making it totally clear to the public that he was focused on jobs and the economy." One DCCC strategist put it bluntly: "We had no economic narrative."

The absence of a compelling and convincing economic message would be a central problem in the months ahead. A vibrant recovery had yet to arrive—and Obama and the Democrats were in charge. A but-for-us-things-could-have-been-worse platform would be difficult to campaign on, however true that might be. Obama needed to persuade voters that he was engaged in concrete actions to revive the economy and create jobs.

Obama did have ideas—such as a proposed $30 billion small business lending program he had promoted at a Buffalo manufacturing plant hours before flying to New York City for that fund-raiser. But the White House did not have an aggressive create-jobs-now blueprint to tout. The administration had even engaged in a fierce debate about whether to concoct such a plan.

The week before Obama's speech, at a private meeting with liberal advocates in Washington, Larry Summers, the director of the president's National Economic Council, was asked why Obama was not speaking more boldly and clearly about job creation.

Summers, the former Harvard University president who had served as Treasury secretary during the Clinton years, was seen as imperious by many inside and outside the White House, and progressive policy advocates in town considered him and Treasury secretary Timothy Geithner a centrist, pro–Wall Street cell in the administration that was blocking more populist economic measures and messages.

Summers had been a cheerleader for financial deregulation in the Clinton years; Geithner had been president of the New York Federal Reserve Bank and a key player in the bailout bonanza that saved the titans of Wall Street who had behaved irresponsibly on his watch. When Obama placed his all-important economic policies in the hands of these two men, it sent a signal that he was receptive to the needs of Big Finance, even while he vowed reform.

Still, Summers and Geithner were not always the pro-corporate obstructionists their critics envisioned. At times, they were stymied by the administration's politicos.

Summers informed the liberal advocates that the White House did not have much in the works regarding new and extensive job-creating initiatives. He and other policy people in the administration believed the government still needed to goose the economy, but he blamed "the political folks" in the White House for shifting toward deficit reduction. Summers's audience drew the conclusion that Obama and his strategists were concerned about polls showing that many voters—especially independents—had absorbed Republican talking points: Obama had spent too much money, the stimulus had not worked, and the government needed to go on a starvation diet.

A few days later, Peter Orszag, the head of the Office of Management and Budget and a fierce deficit hawk, privately told a few of these policy advocates that the White House was definitely pivoting toward deficit reduction. He said that the White House had calculated that any new stimulus proposed by the president would die in the Senate.

According to Orszag, the plan was for the president to act and look serious on the deficit in the near term to accrue political credibility that could later be redeemed to win job-boosting government spending—maybe a modest $100 to $200 billion—in 2011 and 2012, in time for Obama's reelection. (Months earlier, during his first State of the Union address, Obama had talked more about his steps to reduce spending—freezing certain government spending, creating a deficit-reduction commission—than about creating jobs.)

Orszag's message was unambiguous: forget about more

spending to create jobs this year. The Obama White House was tapped out.

THROUGHOUT THE PAST MONTHS, OBAMA'S WHITE HOUSE HAD been conflicted over what to do about the sickly economy. The previous fall, much of the president's economic team had concluded that the recession was worse than they had anticipated.

Yet its members couldn't agree on how to respond. Larry Summers; Christina Romer, the academic who chaired the president's Council of Economic Advisers; and Jared Bernstein, the labor-friendly advocate who was Biden's chief economic aide, were each keen on more stimulus—not another $800 billion, but several hundred billions of dollars.

Orszag, an establishment-oriented economist who had once headed the Congressional Budget Office, was more concerned with the looming deficit. Geithner, who had worked for Summers at Treasury during the Clinton administration, believed that the economy could use another $400 to $500 billion shot in the arm, but he preferred that such a boost be tied to a deficit-reduction strategy. Geithner, though, was often hard to read.

"He holds his cards close," a past colleague said. "You're never quite sure what he's saying to the president or what he believes."

"There was discombobulation," a former senior administration official later complained. "No coherence within the economic team."

It didn't help that Summers and Orszag disliked each other. Or that Geithner believed Summers often blocked policy ideas that did not originate with Larry Summers. Moreover, others in the White House—chief of staff Rahm Emanuel, David Axelrod, and the legislative shop led by Phil Schiliro—had concluded that it would be politically near impossible to do anything big.

"We kept fighting the same fights in front of the president,"

Romer recalled. "He was frustrated about where the economy was, where the politics were, and he was frustrated by the economic team."

When Obama met with his economic aides, Romer would constantly assert that more stimulus was necessary. (She, too, worried about deficits, but didn't want to clamp down until the economy was in a better place.) At one meeting, when Romer was making her now-routine argument for more stimulus, Obama interrupted her: "I can't get it done. Don't you understand that?"

The original stimulus had become a political albatross—rendering the economic team's internal debates almost moot. Moreover, the ever-uphill push for health care reform was consuming most of the available oxygen. Obama had contended that his health care overhaul was essential to improving the jobs picture—and he had strong policy arguments for that case. Yet when the legislation was passed, it was not regarded as a jobs initiative.

"The bill was hard to comprehend," a top administration official later said. "The process was hard to follow. Voters saw him fiddling, while jobs were not being created."

ON APRIL 20, 2010, AN EXPLOSION OCCURRED ON THE BP-licensed Deepwater Horizon drilling rig in the Gulf of Mexico, and the president and his aides were overwhelmed by the worst environmental disaster in decades.

The BP oil spill added to the already present sense of crisis. A Greek financial crunch was roiling markets and confidence around the world. Stock prices were falling. The forecasts from earlier in the year predicting the US economy would rebound were not holding up. And as the oil gushed into the sea, the president could do nothing to stop it. He asked Defense secretary

Robert Gates and Admiral Michael Mullen, the chairman of the Joint Chiefs of Staff, if the US government had any technology that could be used—maybe a secret submarine—to plug this hole in the seafloor. The answer: no. Robert Gibbs recalled that the spill "gave people a sense of futility."

"This was the worst possible crisis for us," Ron Klain, Biden's then chief of staff, subsequently said. "Our own progressive allies were criticizing our response and driving the idea of government incompetence, which would only lead to less support for government action on the economy: 'Hey, if you can't clean up the oil, how can you spend hundreds of billions of dollars to create jobs?'"

White House aides were quite touchy on this front. They believed that criticism of Obama from the left or the right eroded public faith in the president and in government generally, making it more difficult for the administration to propose solutions.

"If we really had given too much money to the banks in TARP funds, cozied up to drug companies with health care reform, and were in bed with BP, why would you trust us with hundreds of billions of dollars?" a former senior administration official later said.

And if the critics on the left were claiming that over $800 billion of stimulus hadn't worked, that made it hard to argue that more money should be spent on similar measures.

Throughout the spring, the president pored over the economic data routinely dumped on his desk, looking for positive signs. "Maybe this is just about to get better," he told aides. The president, Gibbs recalled, was hoping to catch a break on something, anything—just some good news.

"But whatever scenarios we envisioned," Gibbs said, "it always turned out to be a little worse."

IN JUNE, BIDEN LAUNCHED "RECOVERY SUMMER," A SIX-WEEK-long tour to highlight projects made possible by the stimulus. The vice president was optimistic that by pointing to examples of stimulus dollars being spent on productive projects—roads, bridges, and the rest—the administration could shove public opinion in a positive direction. The tour never gained traction.

"The summer didn't feel very recovery-ish," a Biden aide subsequently observed. A similar effort to promote the health care reform act also fizzled.

"The trick was to make it clear that Obama's main focus was still jobs and the economy," one aide said, "when there was so much to deal with—the firing of General McChrystal [the top US commander in Afghanistan who had insulted administration officials in a *Rolling Stone* interview], the oil spill. And in the states, Democratic candidates were getting the shit knocked out of them on 'Obamacare' and deficits by Republican and conservative outside groups spending tens of millions of dollars."

A former senior White House official recalled, "The oil spill was the moment when we realized we had no chance in the November elections. The world seemed going to shit. The president was doing his best, but it was impossible to get credit."

IN MID-JULY, CONGRESS FINALLY APPROVED THE WALL STREET reform bill that the president had been championing. The law rewrote rules governing financial institutions to address systemic risks and lessen the odds that a future financial meltdown could trigger another recession. Wall Street critics griped that the legislation did not go far enough, while Republicans shrieked that it would handcuff financial firms and cause them to lose a competitive edge in the international marketplace.

A prominent provision set up an independent office called

the Consumer Financial Protection Bureau within the Federal Reserve; its mandate was to prevent financial firms such as banks, credit card companies, and mortgage lenders from exploiting consumers with tricks and traps. But the overall legislation was complicated, addressing a complex set of problems—and not easy for voters to sort out.

Again, Obama did not score much political profit from this victory—possibly because he never fully decided whether to blame or court Wall Street. He had vacillated between slamming Big Finance for causing the US economy to collapse while also trying to win over its leaders. In pressing for the Wall Street reform bill, Obama had rarely called out the specific financial firms that had unleashed an army of lobbyists to stop or weaken the measure.

White House aides frequently told Obama that there was great antipathy within the business community toward him. The president and his advisers considered this irrational. Hadn't Obama saved the economy from going over the cliff? He had supported the vastly unpopular bailout of financial firms. He had rescued the about-to-collapse auto industry.

"It wasn't lost on us that stabilizing the major financial institutions at the heart of the crisis was the worst politics in the world," Gene Sperling, a senior economic adviser, later said. "We took the steps we did because we had absolutely no other choice if we were going to prevent a serious risk of another meltdown that would have harmed tens of millions of working families."

Obama had occasionally tossed off a dash of populist rhetoric. Once—just once—he publicly referred to "fat-cat bankers on Wall Street," while denouncing multimillion-dollar bonuses given to financial executives, and he had passionately supported the CFPB, which the financial community bitterly opposed, especially given the possibility that Elizabeth Warren, the populist

Harvard law professor who cooked up the idea for this watchdog office, might become its first chief. (Obama would tap her to set up the CFPB but not to run it.)

But the president had not urged Americans to grab their pitchforks and march on Wall Street. He had disappointed progressives by not making a sustained effort to capitalize on populist anger. Yet for all his reasonableness, business leaders considered him the enemy.

Summers thought Obama needed to choose a side: either blast Wall Street consistently (and capture that populist energy) or reach something of an accommodation with the ringleaders of finance and industry. Not surprisingly, Summers favored the latter. He believed that if those in the business community were confident and prepared to invest more, that would help generate economic growth—the name of the game for him. (Naturally, Geithner was not a champion of revved-up financial regulation; he wasn't a fan of the CFPB and was widely regarded as hostile toward Warren.)

Summers feared that Obama was like a person on a boat who runs from starboard to port and back again, placing the vessel in jeopardy of tipping. The president would assail Wall Street—enough to tick off business leaders but not enough to register deeply with voters. Then he'd reach out to business leaders and meet with them. Later those same leaders would complain to White House aides that these sessions were counterproductive, that they felt Obama had lectured them, and that when he asked them for suggestions, there was not much follow-up from the White House.

Obama and his aides weren't worried that wrathful corporate leaders would take their hefty political contributions elsewhere. They were concerned that constant carping from CEOs would influence how independent voters perceived the president.

A steady stream of corporate grousing about the president—

even if Wall Street had screwed the rest of the nation—could cause these Americans to question whether Obama really knew what he was doing concerning the economy. Obama and his aides fretted that such a perception could lessen his chances of winning reelection.

AS THE YEAR PROGRESSED, OBAMA HAD TO CONSIDER WHAT TO do about the Bush tax cuts. As a candidate, he had vowed—no ifs, ands, or buts—to end the Bush tax breaks for the rich. But there was a problem. Obama had also pledged not to raise taxes on people who pulled in less than $250,000.

If all the Bush tax cuts expired at the end of the year, middle-class Americans would see their tax bills expand. To keep both pledges, Obama would have to detach the top-bracket tax breaks from the middle-class tax cuts and pass legislation to continue only the latter. Republicans could be expected to holler and do whatever they could to block this.

In late July, Geithner, who considered the Bush tax cuts for the rich an impediment to fiscal balance, hit the Sunday morning talk shows and declared that the administration favored letting the tax breaks for the wealthy perish. He dismissed the GOP arguments that allowing tax rates on the wealthy to rise would impede economic growth.

A week later, Geithner again ripped into the Bush tax cuts: "There is no credible argument to be made that the purpose of government is to borrow from future generations of Americans to finance an extension of tax cuts for the top two percent."

He called it a weak way to stimulate the economy and "a $700 billion fiscal mistake." He batted away the GOP contention that ending these tax cuts would be bad for small businesses, noting that only 3 percent of small businesses would be affected by the expiration of these tax cuts.

But Obama and others at the White House were not interested in having this battle yet.

"There was no evidence that the White House wanted this fight now," said an outside Obama adviser. "Tim and his people were saying, 'Let's set this up and have a vote.' But there was no traction in the White House. He went out and laid the groundwork. Then there was nothing."

Larry Summers, for one, "didn't think there was a voter jihad in America against high-income tax cuts and was skeptical of a political strategy to have the defining issue be about high-income tax cuts," according to a former senior administration official.

Senate majority leader Harry Reid and House Speaker Nancy Pelosi each favored killing the upper-income tax cuts and continuing the middle-class tax cuts. But the Democratic caucuses in both chambers were not fully behind their leaders. With the grim midterm elections nearing, some nervous Democrats feared the Republicans would spend gobs of money on ads bashing them as tax hikers, even if they'd only said no to extending tax breaks for millionaires.

"We essentially reached the conclusion that it would be best if each individual senator fought this battle on their own terms," a senior Democratic Senate aide said.

In other words, it was every man and woman for him- or herself.

Democrats on the Hill were griping that the White House was not effectively defending Obama's actions, and there was no coordinated position on the tax question or a big message on the economy. "We just lacked a real battle plan heading toward the election," a top White House official recalled.

IN AMERICAN POLITICS, AUGUST CAN BE A CRUEL AND SILLY month. A year earlier, in the summer of 2009, the Tea Party had waged an angry crusade against Obama's health care initiative, advancing the false charge that it would establish death panels.

Summer 2010 saw a cable TV–fueled dustup focused on what critics were calling the "Ground Zero mosque"—an Islamic community center planned for Lower Manhattan, blocks away from where the Twin Towers had once stood. And Obama fell into the controversy.

At a White House *iftar* dinner on August 13, the president said he believed Muslims have "the right to build a place of worship and a community center on private property in Lower Manhattan, in accordance with local laws and ordinances."

This remark led media outlets to report that Obama supported the Cordoba House project.

Then the next day, Obama explained, "I was not commenting, and I will not comment on the wisdom of making the decision to put a mosque there. I was commenting very specifically on the right that people have that dates back to our founding."

Nuance noted—but news reports treated Obama's clarification as a retreat from the previous night's statement. Fox News reported that Obama was "under fire" for his remarks.

"One of the things that got the president annoyed was frivolous cable TV debates," Bill Burton recalled.

But that summer, Gibbs noted, "If people turned on the television and saw stories and shouting about BP and the mosque, they'd be right to ask, 'Is anybody paying attention to what's important in my life?'"

IN MID-AUGUST, OBAMA AND HIS ADVISERS FACED A NEW SET OF worries when a fresh crop of economic indicators made them wonder if the recovery, fragile as it was, was stalling out. On a family vacation in Martha's Vineyard, Obama held a conference call with his economic team.

We need new ideas that will generate jobs, he said, and he pressed them to devise options quickly. Obama's aides were not optimistic that they could produce a practical jobs strategy that could win approval in Congress.

"What else could we do?" a senior administration official later recalled. "Crisis, war, the BP spill, and health care had already sucked up all the oxygen. We had been trying for months to find things. We kept getting overwhelmed. We tried for nine months to get a small business loan bill passed. And it hadn't yet passed. That was outrageous. Once you lost the sixty votes in the Senate"—which happened the previous January when Republican Scott Brown won a special election to fill Ted Kennedy's seat—"it was tough."

Summers was in charge of conveying ideas to the president. His lack of people-pleasing skills was hardly a secret, and he was regarded by some colleagues in the administration as an obstacle to policy progress.

"The NEC was the place where ideas go to die," quipped a senior administration official who worked closely with Summers. Another aide said, "Larry was good at recognizing ideas—the ones he agreed with."

As NEC director, Summers was supposed to be an honest broker. Yet he was a fellow who knew—or thought he knew—what made a good idea or a bad one, and he didn't mind applying his own judgment.

He saw the need for more stimulus. But he disliked programs that sought to boost certain sectors—say, financing clean energy or high-speed rail. He didn't believe such policies were effective. The Treasury Department sent over proposals to assist small businesses. Summers didn't think they passed muster.

President Bill Clinton kept pitching an idea to Obama and others in the administration: use government bonds or loans to underwrite a massive rehabilitation of commercial real estate to render buildings more energy efficient. This was based on a retro-fitting project at the Empire State Building. The former president had raised this idea so many times with administration officials that it had practically become a joke.

"Every time Clinton saw Obama, he'd press that," recalled one senior administration official. "Obama would then ask Larry about it, and Larry would come back with fifty reasons not to do it."

Summers had contacted people who worked on the Empire State Building project and concluded it could not easily be repli-cated. Summers saw himself as being rigorous—at a time when there was little margin for error. Yet others would point out that while Summers was nay-saying the ideas that flowed in, he wasn't devising killer proposals of his own.

Christina Romer recalled that August 2010 was spent fran-tically "trying to come up with ideas." In meetings, Obama's economic advisers would envision a Venn diagram—one circle would contain policy prescriptions Democrats supported, the other had ideas Republicans favored—and they would concen-trate on the intersection.

"It was vanishingly small," Jared Bernstein recalled: infra-structure, small business assistance, tax cuts. "Job creation is a tough thing to do when the federal government is not willing to go all WPA-y, when we're not creating direct jobs, not hiring thousands of people to build a dam," Bernstein added.

At the same time, political aides kept telling the economic aides that voters didn't want more spending. As a former senior administration official later noted, "The political people were saying, 'Please don't give us any more policy. Your policy is kill-ing us.'"

"There was literally nothing you could do in the summer of 2010 that would create a job by November—even if we passed $400 billion of stimulus," Ron Klain later said. "The policy people were stuck on a sofa scratching their heads. To make matters worse, John Maynard Keynes had been clubbed to death. No one wanted to stand up and say let's spend more money."

AFTER OBAMA RETURNED TO THE WHITE HOUSE, HE DELIVered a prime-time speech announcing the end of the combat mission in Iraq and held a series of meetings with Middle East leaders to jump-start yet another peace process.

But on his first day back, he held a meeting with his economic team. More ideas, he urged them, more ideas.

Afterward, in the Rose Garden, he said: "My economic team is hard at work in identifying additional measures that could make a difference in both promoting growth and hiring in the short term, and increasing our economy's competitiveness in the long term."

But he had no details to offer. Obama also asked the Senate Republicans "to drop the blockade" against legislation that would provide loans and tax breaks to small businesses to encourage them to hire workers; it had been languishing in the Senate for months. Many Democrats thought it was too mild a rebuke, considering that the election season was starting.

The president and members of his economic team knew they needed to demonstrate their commitment to jobs, jobs, jobs, but they believed they had not received sufficient credit for having passed a variety of modest-sized jobs measures over the past year: cash for clunkers, expanded unemployment benefits, a tax credit for new home buyers, a tax credit for hiring, and a jolt of fiscal relief to the states that saved the jobs of thousands of teachers.

That might be so, but as the final leg of the 2010 campaign was beginning, the issue was not just policy realities but political image. If an unemployed person in Ohio were asked whether she believed that Obama and his crew were working 24/7 to revive the economy, what would the answer be?

In early September, the president hit the road—in sort of a kickoff of the fall campaign season. At Cuyahoga Community College, near Cleveland, Obama assailed the Republicans for cutting taxes for millionaires; for cutting regulations to help special interests; for cutting trade deals that didn't benefit American workers; and for cutting back on investments in education, clean energy, research, and technology.

Their idea, he said scornfully, "was that if we just had blind faith in the market, if we let corporations play by their own rules, if we left everyone else to fend for themselves, that America would grow and America would prosper."

All this, he said, had led to a weaker economy and falling wages—and a financial crisis that triggered the "worst recession of our lifetimes."

He accused the Republicans of trafficking in fear and offering no more than the failed policies of the past.

Obama derided Representative John Boehner, the House Republican leader, for blocking that small business jobs bill backed by the Chamber of Commerce—"They're making the same calculation they made just before my inauguration: if I fail, they win."

He slammed Boehner for opposing other modest economic initiatives proposed by the White House: a boost in infrastructure spending, a research-and-development tax credit, and more generous provisions for businesses to write off investments in plants and facilities.

Obama noted that he wanted to make the middle-class tax cuts permanent, while the Republicans insisted on giving millionaires an extra $100,000 a year—which would force the US government to borrow another $700 billion to cover this revenue loss.

"We should not hold middle-class tax cuts hostage any longer," he proclaimed.

Obama said little about the reckless Wall Streeters. He vowed to reduce the deficit. And he stuck with the frame he had first deployed the previous May at that fund-raising dinner: "Do we return to the same failed policies that ran our economy into a ditch, or do we keep moving forward with policies that are slowly pulling us out?"

He had hit all the obvious points, but he did not tap into the anger of the voting public. This was certainly an unconventional political moment, with unemployment so persistent and so high. Could conventional political strategizing prevent a Democratic wipeout at the polls? Perhaps not much could be done at this stage. Nevertheless, congressional Democrats were worrying that Obama was not punching Republicans hard enough or often enough.

MEETINGS WERE HELD IN THE OVAL OFFICE AND ELSEWHERE IN the White House, where Obama and his aides continued to hash out what to do about the expiring Bush tax cuts. Just about everyone favored undoing the high-income tax breaks, but Larry Summers voiced concern that such a battle could negatively affect business-class attitudes toward the president. As one senior official put it, Summers "really hated being accused of engaging in class warfare."

David Axelrod contended that the Democrats on the Hill had the politics wrong and that a vote to end the top-bracket tax cuts would help them in the election.

But vulnerable House and Senate Democrats were still insisting that they did not want to take such a vote. Vice President

Biden was fond of saying—during these deliberations and at other times—"You should never tell a politician you know his interests better than he knows his interests."

"The polling clearly showed that the American people didn't buy the argument that if you make $250,000 you shouldn't pay more," a senior administration official later said. "It was like a 70–30 spread in the polls. But we couldn't translate this into votes on the Hill. You could show legislators this data, but they were convinced that the American people would see them as tax raisers."

The Supreme Court's *Citizens United* decision, rendered earlier in the year, had unleashed outside groups to spend untold amounts of money—some of it from secret sources—against candidates, and pro-Republican outfits could be expected to exploit this decision and hurl millions of advertising dollars accusing Democrats of raising taxes.

In light of this, Obama told his aides there was not much he could do to persuade congressional Democrats: "These guys are the ones on the ballot. They'll decide what's best for them."

He would not lead a pre-midterms fight on his own.

There was an upside to the Democrats' internal conflict over the Bush tax cuts. Obama's economic advisers started wondering if the president could use the tax-cut issue to advance other economic policies that Republicans would otherwise filibuster and kill in the Senate. After all, to extend the Bush tax cuts (for high-, middle-, and low-income earners), Congress would have to pass legislation.

The question became, could Obama graft a new bout of stimulus onto tax-cut legislation? according to a senior administration official.

Geithner, Summers, and others began considering what the

White House could obtain in return for acquiescing to political reality. Extended unemployment benefits? Infrastructure spending? An additional middle-class tax cut that would pump up consumer demand? Obama and his aides were not, as they saw it, retreating on his campaign promise to bury Bush's high-end tax cuts. They were putting a crisis to good use.

This strategic shift was not advertised. Most of the political world saw a brewing showdown over whether or not Obama would succeed in killing the upper-income tax cuts. But to Obama, it was not about winning the fight; it was about exploiting the fight.

"This was the policy choice: how to smuggle in more stimulus," one Obama adviser later said. "And nobody—I mean nobody—in Congress was focused on this."

WITH THE MIDTERM ELECTIONS UNDER WAY, THE STRATEGISTS at the DCCC believed that Obama was doing what he could for them in terms of raising money—no small task. But House Democrats, fearing the worst was hurtling their way, were exasperated with the president.

"There was an overarching frustration," a top House Democrat recalled, "that the administration had not done a better job explaining what it had done—and what we had done—going back to the stimulus, to everything through the health care bill. We were going into the election almost defenseless, and susceptible to a serious charge: you're not looking out for the economic interests of the country."

At the White House, though, Obama and his aides were not convinced that nationalizing the election—having the president make campaign stops across the country to bash the GOP daily—would change anything.

"Everyone had hoped that the recovery would be stronger and

faster," a White House aide said. "Now we realized it wouldn't confine itself to the campaign schedule."

Obama, too, was irritated. In a meeting with a key adviser, he asked, "Why was it so much easier during the campaign to drive the message? We did it every day and it got through." Obama felt he had saved the country from an economic depression, rescued the all-important auto industry, changed the rules for Wall Street to prevent another financial crash, enacted historic legislation reforming the health care system. Yet for all that, he and his party were still struggling.

"With all due respect," the adviser replied, "you're president of the United States. You have wars, oil spills, financial crises, pandemics—you have to respond to all of that. We used to just have to decide what speech to give and what ads to take out. Now you don't have the clarity and simplicity of a campaign message. That's the nature of the beast, especially at a time of abnormal events."

"Yeah, but . . . ," Obama said. "I hear all that. But we have to do better."

OBAMA WASN'T CAMPAIGNING WITH CANDIDATES IN CONGRESsional districts. Some Democrats complained about that, yet many others weren't yearning for his presence. But he was stumping and decrying the Republicans. The president and his political strategists still believed that a large part of Obama's appeal for independent and moderate voters was his reluctance to act in an overly partisan fashion. Having Obama slamming Republicans, one aide noted, "was damaging to our political brand. But he was willing to do this to help the Democrats. We knew this would hurt his numbers with independents."

As election day neared, pollsters working with Democratic House incumbents told their clients that the forward-not-backward/in-the-ditch-or-out-of-the-ditch message had run its course and was no longer effective—especially with independents. Voters in many districts didn't seem afraid to hand the keys to the car back to the Republicans.

The White House recalibrated its campaign message and began accusing the Republicans of being handmaidens of corporate special interests that were trying to steal the election by pouring huge amounts of secret money into races across the country.

Thanks to the *Citizens United* ruling, outside groups, including those set up by millionaires and industry associations, were indeed flooding congressional districts with millions of dollars' worth of negative ads sliming candidates, and the voters couldn't always tell who was behind the assault.

The president publicly decried this "flood of deceptive attack ads sponsored by special interests using front groups with misleading names." Obama pointed out that he and Democrats had tried to pass a law that would force the sponsors of political ads to identify themselves and the sources of their funds—but Republicans had blocked the bill.

The president and his crew were promoting the specter of special interests plotting with a partisan minority in Congress to mount a "power grab" because Obama and the Democrats had dared to defy them with health care reform, Wall Street reform, and other measures. But top Democratic strategists working with the DCCC were not convinced the White House plan would work.

"I don't know any professional politician who believes we could win an election by accusing the other side of spending too much money," one complained after the elections. "That was just background noise to the economy. And we were engaged in some of the same stuff."

This debate played out in an otherwise routine meeting at the White House on October 8. The communications office had called in a squad of Democratic strategists. Not for advice. The point was to share with these "talkers"—who appeared on the cable TV circuit—the White House's strategy for the final weeks of the elections.

The White House held a meeting like this every few months and invited some version of this group, which included Dee Dee Myers, Peter Fenn, Simon Rosenberg, Kiki McLean, Jamal Simmons, and others.

Obama dropped by the meeting. He tried to shore up the partisan loyalists. He pointed out that Bill Clinton had done worse in the polls at this stage of his presidency, and unemployment in 1994 had been only 5.5 percent, not the current 9 percent. Noting that the Republicans and their allies were outspending the Democrats, he said, If it were a fair fight, we'd have better odds.

"It was more of an excuse than a way to address the situation," one participant recalled.

After the president departed, David Axelrod said the remaining weeks would focus on one dramatic story line: the Republicans were stealing the elections with mystery groups spending too much secret special interest cash.

Several of the strategists agreed with this course, some said nothing, and others suggested ways to tweak the message. When it was consultant Chris Kofinis's turn to weigh in, he let loose. At the time, Kofinis was advising Democratic congressional candidates around the nation. Every poll he had examined told him that voters' number one issue was the economy—particularly, the absence of jobs.

"David, with all due respect," Kofinis said, "this message won't work."

The room went silent. Kofinis continued: if the president and the White House hit the campaign trail assailing *Citizens United,* special interests, and money in politics, it wouldn't resonate with voters; most people were unfamiliar with *Citizens United* and more concerned about the economy. Kofinis could not conceive of a more disconnected message for Democrats to take to the voters.

Axelrod countered that voters would respond to a Democratic attack on oil companies and other corporations that were buying influence and tilting the political system in their favor.

Kofinis understood the pernicious influence of corporate money but he insisted that voters were not fixated on such matters: "They're focused on jobs."

He was worried that if the White House and Democrats led with this argument, they would come across as out of touch. Voters dealing with tough economic times would come to believe, he said at the meeting, "we don't get it."

"This conversation wasn't going to change anything," one participant later said. "Everybody knew this wasn't really working. But the White House was not going to rip up their strategy at this point. We all were just waiting for the storm to arrive."

Two days later, Axelrod appeared on *Face the Nation*. That morning the Democratic Party had released an ad accusing Republican strategist Karl Rove and former GOP chairman Ed Gillespie ("Bush cronies") and the Chamber of Commerce ("shills for big business") of "stealing our democracy" by spending "millions from secret donors to elect Republicans to do their bidding in Congress."

The spot showed a shadowy figure mugging a woman and absconding with her purse, and it ominously proclaimed, "It appears they've even taken secret foreign money to influence our elections."

But the previous day, the *New York Times* had reported that the Chamber did not receive significant sums from its foreign affiliates, and moderator Bob Schieffer challenged Axelrod to back up this dramatic foreign money claim.

Axelrod couldn't, but he noted that the existing rules prevented anyone from determining the source of the money for the Chamber's massive campaign against Democratic House members. That was true, but the White House had botched its message. By zeroing in on the foreign money angle—which it couldn't prove—it had served up a refutable charge that distracted from its main point: the GOP was in league with corporate interests to rig Congress in the favor of industry.

With three weeks to go before the election, is this the best Obama and the Democrats could do? Schieffer asked. To charge that there "may or may not be foreign money coming into the campaign"?

Axelrod reiterated the "fundamental" concern: Republicans and their backers "want to turn the clock back to the very same policies that got us into this mess in the first place, that exploded our deficits, that put the special interests in control . . . the oil industry, Wall Street, insurance industry."

This was an accurate description, but Axelrod said nothing about the president's achievements or any Democratic plans for dealing with the still-flagging economy. He was talking about the "threat to democracy," not about putting people back to work. The two were linked—corporate-connected GOPers had blocked Obama initiatives to bolster the economy—but the White House was playing a bank shot.

Democrats felt they couldn't run on their significant legislative accomplishments.

"That left us with Republican ethics," a DCCC official said. "And that led to us not having a dialogue with voters about what they care about. There was no jobs message. The voters, rightly or not, saw debt as a contributing factor to the bad economy, and we were talking about who was spending what money in politics. We had an election driven by the enemy. Their message was simple: the Democrats are spending too much and it's hurting the economy. There was no economic narrative coming out of the White House. Then again, perhaps no messaging tweaking could have affected what was about to happen."

"We lost all hope in October," a top White House official recalled. "We didn't feel much of anything. We just had to let the ship hit the iceberg."

ON ELECTION NIGHT, MANY WHITE HOUSE AIDES WENT HOME early, and the president headed off to the residence.

Chapter Two

THE DEAL THAT WORKED

THE MORNING AFTER THE MIDTERM ELECTIONS, OBAMA HELD A meeting with his senior staff. He had just been hit with a historic loss. "Everybody was prepared for the worst—and it got worse," Robert Gibbs recalled. "We'd be playing defense for the next two years. People were very down and depressed." Obama's presidency was on the rocks.

As David Axelrod later put it, "Everyone was hanging crepe." Not Obama. This is not the time to gnash teeth, he told his aides, we have a lot of stuff to get done. The Democrats still controlled both houses of Congress through the end of the year.

He ticked off a list of what he aimed to achieve in the next two months: a tax deal, extending unemployment benefits, ratification of the New START treaty reducing nuclear arms, repeal of the Pentagon's Don't Ask/Don't Tell policy preventing gays and lesbians from openly serving in the military, passage of the DREAM Act (which would grant citizenship to undocumented young adults who met certain requirements), and a children's nutrition bill advocated by Michelle Obama.

"One of these alone was a heavy lift," Axelrod said. "Together, it seemed a leap of faith. We all looked at each other quizzically. 'What does he see that we don't?' "

THAT DAY, THE PRESIDENT HELD A PRESS CONFERENCE. HE DID not come out swinging; he was defensive. He said he understood if some voters, in response to the stimulus and the bank and auto bailouts, "felt as if government was getting much more intrusive into people's lives than they were accustomed to."

He added, "I'm sympathetic to folks who looked at it and said this is looking like potential overreach."

Despite having ridden to victory in 2008 on the theme of change, he had not been able to transform Washington, but he felt he had a good reason: "We were in such a hurry to get things done that we didn't change *how* things got done."

His spin on the election was that voters "want everybody to act responsibly in Washington" and "work harder to arrive at consensus."

But was that why dozens of Tea Partiers had just been elected—to make politics in the capital more mature?

"I do believe there is hope for civility," Obama remarked.

The same day, Mitch McConnell, minority leader in the Senate, adopted a combative tone: "Our primary legislative goals are to repeal and replace the health spending bill, to end the bailouts, cut spending and shrink the size and scope of government. The only way to do all these things is to put someone in the White House who won't veto any of these things."

He was signaling the essence of the Republicans' hyperpartisan strategy: destroy Obama.

The president had conveyed a much different message: in the coming food fight, he would be the adult in the room, the reasonable man looking for productive discourse and sensible compromise.

Axelrod, according to another Obama adviser, believed the election results meant that independent voters wanted a leader who would make all the squabbling schoolchildren in Washington do their assignments. The president and his message man were betting that such an approach would both win back those in-the-middle voters (even if it frustrated progressive activists and congressional Democrats) and yield policy advances that would benefit the nation.

DURING THE FOLLOWING WEEKS, THE WHITE HOUSE STAFF WAS consumed with a postelection reorganization. White House people wondered—or worried—about who would be departing and who would be getting which jobs. The makeover was being overseen by Pete Rouse, the low-key interim chief of staff. A month before the election, Rahm Emanuel had left his post— abandoned it, grumbled some White House staffers—to run for mayor of Chicago. Obama had tapped Rouse, the methodical deputy chief of staff, to replace Emanuel, on a temporary basis.

Known for his analytical talents, Rouse was a fan of developing strategic plans. After the roller-coasterish first two years of the presidency, Obama could use some strategy. And Rouse had a lot to consider—as Obama and his advisers realized that plenty of missteps, large and small, had led to this low point.

There was no shortage of advice. Anita Dunn, the former communications director, and others believed the White House should boldly define Obama's program and demonstrate he was willing to fight for it. Yet there was another school of thought within the president's circle: Obama's personal attributes mattered most. Independent voters perceived him as too partisan and too liberal—and that was the problem.

This discussion tracked back to internal deliberations held during the 2008 campaign. Should the campaign push Obama himself as the postpartisan change he championed, or did it have to create a mandate for the specific policy proposals Obama would press for once in the White House? Boiled down, the question was Obama the brand or Obama the program. There certainly was overlap. But after the midterm elections, Obama's political aides were focused on repairing the brand.

OBAMA HELD PRIVATE DINNERS AND MEETINGS WITH WASHINGton veterans and outsiders: Leon Panetta, a former Clinton chief of staff now heading the CIA; Kenneth Duberstein, a Reagan chief of staff; former Senator Tom Daschle; David Gergen, a centrist presidential-adviser-turned-pundit; Matthew Dowd, a onetime George W. Bush adviser; John Podesta, who headed Obama's presidential transition and now ran the liberal Center for American Progress; Bill Clinton; and others.

Vernon Jordan, the preeminent Washington fixer, told Obama he had been overly partisan. Clinton administration veterans suggested that Obama be . . . well, more like Bill Clinton.

Some of these interlocutors reiterated the widespread gripe that the Obama White House was too insular. Obama was told he should get out more, strengthen his relationships with other Washington power players, be less aloof from the capital's permanent establishment. He should put the bully pulpit to better use. He should have a more disciplined White House.

Rouse's review was supposed to address the widespread criticism in Washington that the crew at 1600 Pennsylvania was indeed too locked within its own bubble. But the immediate impact of all this assessing of White House process was uncertain.

"Obama didn't care about the criticism that he was too insular," a White House aide said. "He didn't give a shit."

One outside adviser who met with Obama shortly after the

election found that the president was surprised by the depth of the loss. What stung Obama was not the virulent animosity of the Tea Party set, but that he had lost the middle. He believed that one reason for this was that the business community had turned against him so sharply. If the antagonism of the corporate class was causing independent voters to view the president as out of touch with the private sector, the adviser told the president, this would somehow have to be addressed before the next election.

Obama acknowledged that, but he had a more visceral response: "I saved these guys when the economy was falling off a cliff. Now I get nothing but their venom."

The president told another adviser that the elections caused him to conclude that he had to find a way to create jobs and not let anything hold up the recovery. Everything else depended on people believing and feeling that the economy was moving in the right direction. This was hardly a radical observation. Putting it into practice would be the challenge.

ON VETERANS DAY, A DOZEN OR SO PROGRESSIVE POLICY ADVOCATES filed into the Roosevelt Room of the White House for a confab with Obama's top advisers. Larry Summers, Gene Sperling, David Axelrod, Phil Schiliro, and Jim Messina, the White House deputy chief of staff, were present. The visitors included Robert Greenstein, the president of the Center on Budget and Policy Priorities; Bill Samuel, the legislative director of the AFL-CIO; and John Podesta. The subject at hand was the issue that the politerati expected to define this lame-duck session: the Bush tax cuts.

The advocates were nervous about what Obama might do. The issue, they believed, was simple: fight or no fight? But Axelrod reminded them that Obama had promised to end the Bush tax breaks for the rich, but he had also vowed to prevent a tax hike for the middle class.

With the Republicans in the Senate blocking legislation that would extend only the middle-class tax cuts, Obama would have a difficult time honoring both promises. He might have to choose. Thus, the White House was considering the possibility of a deal, instead of a knockdown slugfest.

Summers was blunt: to extend soon-to-expire unemployment benefits, we're going to have to give up on the tax cuts for the rich.

"Getting more for our people is more important than getting less for their people," he said.

Summers explained that these benefits and all the tax cuts would run out at the end of the year and that would place a drag on the economy at precisely the wrong moment. He and the president didn't want another negative jolt to the already anemic recovery. It was far more important to procure more stimulus for 2011 and 2012 than to bicker over the rich receiving unnecessary and deficit-causing tax cuts.

The advocates understood this logic. Still, several maintained that Obama ought to force the Republicans first to vote on the tax cuts, if only to show the world that the GOP was willing to kill middle-class tax breaks to serve the wealthy.

Obama, though, wasn't spoiling for a fight—in part because he didn't trust his comrades. McConnell, he believed, was willing to play a game of chicken and permit all the Bush tax cuts to end on December 31. Then the new House—under the control of Republicans—would quickly pass legislation to renew the tax cuts for rich and middle-income earners and perhaps even make them permanent. The pressure would mount on the Democratic-controlled Senate to do the same. White House officials believed Democrats in that body would eventually blink and agree to an across-the-board extension.

"We're not going to win the battle," Summers told the group, "so let's focus on what we can gain in return."

MANY HOUSE DEMOCRATS WERE PRESSING THE WHITE HOUSE to mount a battle royal. They had just experienced a brutal election. They knew that extending the tax cuts for the rich would create more debt and that would lead to greater pressure for more budget cuts. They at least wanted a public debate that would frame the discussion in their favor.

"We heard Democrats say, 'Let's make them vote over and over again on tax cuts for the rich,'" Axelrod later said. "'Go until January or February and people will know that's what the Republicans stand for.' We were flabbergasted. They missed the overall point that taxes would go up and unemployment insurance would be lost by two million. Obama was determined to get something done."

Especially when unemployment was near 10 percent.

Obama and his aides didn't believe they would gain much politically from a dramatic face-off with the Republicans during the holiday season when it likely wouldn't register much with the American public.

Another compelling reason for eschewing a grand confrontation was the time it would suck up. A drawn-out tussle over the tax cuts would leave Obama little opportunity to achieve the other parts of his lame-duck agenda: New START ratification, the repeal of Don't Ask/Don't Tell, and passage of the DREAM Act.

Obama saw the trade-off clearly. He could please those congressional Democrats and progressive allies craving fisticuffs and (at best) end up with some modest tax-cut compromise and nothing else—or he could quickly negotiate a more expansive deal and then have a shot at scoring a civil rights victory, securing a nuclear arms control treaty, and obtaining a victory for immigration reform.

Over the Thanksgiving holiday, the president discussed this in a long call with David Plouffe, the no-nonsense strategist who had managed the 2008 campaign.

"They will say you caved," Plouffe told him. "That can't be avoided."

I know, Obama said. But that would be the price that had to be paid.

AT THE NOVEMBER 30 DAILY PRESS BRIEFING, ABC NEWS'S JAKE Tapper, noting that the president was scheduled to head to Hawaii for a holiday trip on December 18, asked Gibbs, "The president thinks that funding the government, passing unemployment insurance extensions, Don't Ask/Don't Tell repeal, the DREAM Act, tax cuts, and START all can be done in the next eighteen days?"

Gibbs had a one-word answer: "Yes."

"Good luck," Tapper replied, with a touch of sarcasm

"Well, thank you," Gibbs said. "You'll have a lot to cover."

THAT DAY, OBAMA AND THE DEMOCRATIC AND REPUBLICAN leaders of the House and Senate launched bipartisan negotiations aimed at reaching a tax-cut deal. The first session of these "six-pack" talks, held in an ornate Senate conference room, had not gotten much beyond vague opening statements.

The media reports focused on whether this ongoing powwow would save the expiring Bush tax cuts—and which portions of those tax breaks would remain. Obama's crafty intention to use this issue to advance stimulative policies was not widely known.

The following morning, Obama gathered his economic team

and top advisers in the Oval Office—Biden, Geithner, Summers, Rouse, Klain, Messina, Jarrett, Schiliro, Lew, and others—to strategize how best to negotiate a tax-cut package fast.

Biden emphasized that if Obama challenged the Republicans on tax cuts, the Senate would bog down and the White House could lose New START. Ratifying the nuclear arms treaty was a top priority for Obama. If it flopped in the Senate, Obama would look weak on the world stage; his attempt to reset relations with Russia (a necessary component of his efforts to isolate Iran and inhibit its nuclear program) would be in tatters. Biden reported that Senate Republicans were telling him the tax-cut issue had to be settled before any other business could proceed.

"This is no bluff," Biden said. And in the next Congress, he warned, ratification might be a total nonstarter.

Obama believed he could win approval of the treaty before the lame-duck session concluded. His questions concerned Don't Ask/Don't Tell.

"Do you really think I could get repeal, if we get the tax-cut deal?" he asked Messina and Schiliro.

They said they believed so, but quickly added there was no guarantee.

Obama told his aides he was prepared to take the heat on a tax-cut deal. He knew liberal Democrats on the Hill and progressives elsewhere would scream. But Obama felt it was worth the political trouble to end the ban on gays and lesbians openly serving in the military.

Let's move quickly, the president told his team.

The six-pack negotiators held two rounds of meetings that day but made no real progress. The Republicans declared they would never agree to any package that decoupled the Bush high-end tax cuts from the others. But these were not the key negotiations.

Weeks earlier, Obama had handed Biden the top-priority

assignment of getting the New START treaty ratified. The vice president had once chaired the Senate Foreign Relations Committee and for years toiled on nuclear arms control issues. No one in the White House had a clearer grasp of the Senate, its procedures, and its members.

Biden had immediately started talking with Mitch McConnell about the treaty. Now, with Obama's approval, Biden expanded his chats with McConnell to cover the possible shape of a tax-cut deal.

The vice president was old school. He took pride in his decent working relationships with Republicans in the Senate, including McConnell. "Joe does the things that Obama doesn't like to do," a Biden friend remarked. "Like negotiating with the Hill. He loves it."

As the six-pack talks dragged on, Biden and McConnell spoke privately several times. The basic parameters emerged: the Democrats would get an extension of unemployment benefits (which Republicans in the Senate had long opposed) and the GOP would pocket those top-bracket tax cuts. Both sides could claim credit for extending the tax-rate reductions for the middle class.

In his conversations with the vice president, McConnell had noted another thorny issue. If the White House wanted much else in the package, the Republicans would insist on deep cuts in the estate tax for the wealthy. In particular, they fancied a proposal being pushed by Republican Senator Jon Kyl and Democratic Senator Blanche Lincoln, which would raise the exemptions for the estate tax from $3.5 million for an individual and $7 million for a couple to, respectively, $5 million and $10 million *and* cut the rate on all estates larger than those from 55 percent to 35 percent.

This would be a tremendous boon for the wealthiest; about sixty-six hundred estates a year would each gain a break of $1.5 million—and $23 billion in tax revenues would be lost.

For weeks, Obama and his economic team had been hashing out what they would press for in a deal. They hoped to bag extensions for the various tax cuts of the Recovery Act—many of which were designed to assist middle-income families or the working poor. They wanted thirteen months of unemployment benefits for out-of-work Americans. (About two million unemployed workers were about to run out of these benefits, and a White House report claimed that curbed household spending due to the termination of unemployment benefits could lead to the loss of six hundred thousand jobs.)

If Obama could get this, he would pump several hundred billions of dollars into the economy. It would be a second stimulus—without being labeled such.

Meanwhile, the House Democrats held a vote on their plan to extend the middle-class tax cuts and kill the upper-income tax cuts. The vote was decisive, 234 to 188 in favor. But this was irrelevant because the real action was taking place in Biden's and McConnell's offices. It was as if the House Democrats were living in a separate world.

At 9:00 a.m. on Saturday, December 4, 2010, Obama and his top advisers convened in the Oval Office for that decisive meeting. McConnell had tendered Biden a specific offer. If the president would accept a two-year extension of all the Bush tax cuts *and* the Lincoln-Kyl estate tax cut, the Republicans would agree to a long stretch of unemployment benefits and an extension of some of the tax cuts for middle- and lower-income Americans that Obama had included in the stimulus bill.

But, McConnell added, these breaks could not be refundable. Republicans tended to despise refundability, deriding it as welfare. (With a refundable tax credit, a taxpayer who qualifies for the credit but whose tax liability is less than the amount of the credit receives the difference as a refund.)

The GOP, McConnell added, would not support continuing renewable energy tax credits. That tax break was too Obama-ish.

Timothy Geithner spoke first. He was often one of the more prudent voices in Oval Office deliberations. But now he was vehement: this offer was not good enough.

Geithner advised Obama to stick to a simple proposition: if, as the Republicans say, it's wrong to let taxes go up during slow economic times, this notion should apply across the board. The president should demand a continuation of all the key tax credits and breaks for middle- and low-income Americans, the renewable energy grants, and greater depreciation for business investment expenses.

Without more from McConnell, Geithner said, this was a lousy deal for Obama. The Treasury secretary was ready to walk. Others in the room were stunned.

The president was unhappy about the estate tax. He suggested pushing back on that. Geithner proposed that the White House insist on everything it wanted in return for accepting a version of the estate tax cut. And he had a particular demand in mind.

Geithner and Sperling had come up with the idea of dumping the Make Work Pay tax credit that had been in the stimulus package and instead pressing for a yearlong suspension of the payroll tax for employees that funds Social Security. For months, Sperling had been consumed with a payroll tax holiday to add kick to the flagging economy. He had compiled a list of Republicans who had supported it in the past. Though there had been concern that older voters might fret about cutting revenues earmarked for Social Security, the economic team agreed it should be a critical part of the deal. (The tax break would not affect Social Security payments.)

The refundable Make Work Pay credit, which provided $400 to individuals and $800 to joint filers, had been a centerpiece of Obama's stimulus package. But most voters never realized they had received this tax cut, because it had been doled out in small increments over the course of the year via decreased withholding on paychecks. Only one out of ten people in a recent poll knew that Obama had cut taxes for most Americans. It had been yet another communications screwup.

In the Oval Office, Geithner and Sperling noted it would be better to win a new tax cut for middle- and low-income Americans than to continue an invisible one. Another sly reason for the swap was that a one-year payroll tax holiday would inject about $110 billion into the economy; an extension of the Make Work Pay tax credit would do the same, but over two years. The new proposal would produce twice the bang. Plus, Sperling figured, it might be possible in a year to win an extension of this break—would the Republicans really say no to that?—and shove yet another $110 billion of stimulus into the economy.

At this meeting, Phil Schiliro raised a concern about the particularly dicey matter of acquiescing on the estate tax provision. This would enrage Capitol Hill Democrats, he predicted. They despised estate tax breaks for the rich and had been scuffling with Republicans over this for years.

Yield on this, Schiliro told the president, and you'll have a firestorm within your own party.

Others in the room disagreed. Total up what the rich people get, they said, total up what the poor people get, and as long as the poor do better, it will be fine.

No, no, no, Schiliro insisted. The numbers won't matter.

Biden proposed telling McConnell that opposing the estate

tax is an article of faith for most Democrats and that opposing refundability is the same for Republicans. Let's trade one for the other.

Obama agreed—to a degree. He did consider the notion of handing the well-to-do another tax break offensive. He certainly couldn't swallow the estate tax provision without refundability.

"Do we have to give so much on the estate tax?" he asked Biden.

Draw a line with McConnell, he instructed the vice president: "Tell them they're not getting that unless I get refundability."

To aides in the room, it seemed as if this could be a deal breaker.

THOUGH BIDEN DID NOT YET HAVE A DEAL WITH McCONNELL, Obama decided it was time to bring the Democratic congressional leaders into the conversation. It was a delicate moment.

The previous day, the *Washington Post* had reported that the White House had been talking privately to McConnell, outside the six-pack process. "The shadow talks," the paper noted, "have stirred considerable ill will among Democrats, who complain privately that the White House is capitulating to Republicans without extracting anything substantial in return."

The story was wrong in that regard, but it fueled the impression that Obama was preparing to fold on the tax cuts for the rich and sell out his party colleagues.

With intraparty apprehension in the air, Obama asked Harry Reid and Nancy Pelosi to come see him at the White House— that day. The two showed up in the afternoon.

In the Oval Office, with Biden and members of the economic team present, Obama described the current state of play to Pelosi and Reid: Joe and Mitch have been talking informally and in the course of these chats, Mitch gave Joe an informal offer, and we're curious as to how Democrats in the House and Senate might react.

Reid, as one participant later recalled, "responded with a very polite fuck-you."

It's clear, he said to Obama, that you and the vice president have been negotiating directly with McConnell, and if you two think you can do business directly with him, feel free to bring whatever you cook up to the Senate floor and see how many votes you can get.

Reid didn't express any anger. His message was more of a dismissive good-luck-it's-not-my-problem.

Pelosi was more sympathetic. I understand why you did this, she said, but my caucus is shell-shocked from the elections and pissed off about . . . well, about everything.

She didn't know how her colleagues would respond to this sort of compromise. She asked if she could bring other House Democratic leaders into the conversation.

Obama and Biden suggested that Pelosi and her allies meet with Biden that evening. When the conversation ended, Obama hugged the speaker, and then placed his hand on Reid's back, as he escorted him out of the Oval Office.

THAT SATURDAY NIGHT, PELOSI ARRIVED AT THE VICE PRESIDEN-tial residence at the US Naval Observatory with Representative Steny Hoyer, the House majority leader, and Representative Chris Van Hollen, who had been participating in the six-pack talks.

Biden had ordered up a fine meal for the occasion. Timothy Geithner, Jack Lew, Pete Rouse, Ron Klain, and John Lawrence, Pelosi's chief of staff, were at the table, too.

As they dined, Biden discussed details of the likely agreement.

Pelosi appeared resigned to the compromise. She told Biden that if you say this is the best you can squeeze out of the Republicans, I believe you, but many of my members won't.

Hoyer responded more positively; Van Hollen was irritated. The White House told us the six-pack talks were real, he protested, but you were negotiating on your own.

Then came dessert. As it was served, Biden revealed that the Republicans would also get the estate tax break. All three House Democrats were upset. Pelosi called this "crazy" and said that such a package would carry no Democratic votes. Hoyer also criticized this decision.

Van Hollen, as one participant later put it, "was on fire." He was dead set against yielding on the estate tax. There was no economic rationale for such low rates on inheritances; they wouldn't spur economic growth. It was just another giveaway to the rich—on top of the Bush tax-cut concession. This was going too far to strike a deal.

The three argued the matter with Biden, and then Geithner spoke up. Biden bristled a bit as Geithner took charge. The Treasury secretary told the House Democrats that he hated this provision as much as they did and perhaps more so than Biden. (The vice president didn't appear to appreciate this quip.)

Geithner continued: I've been in these talks, and I can tell you that what we're getting is worth it. You've been in the majority, and you haven't been able to win extensions of refundable tax cuts for low-income Americans. We're going to lose those tax cuts without a deal. It's ghastly but it's worth it.

The three saw there was not much use quarreling further. Hoyer insisted that a better deal be negotiated. Van Hollen remained upset about the Biden-McConnell negotiations. The White House, as he saw it, was not playing straight.

ON THE SUNDAY MORNING NEWS SHOWS THE NEXT DAY, McConnell and Kyl each hinted that a deal was close: extending all the
Bush tax cuts in exchange for extending jobless benefits for millions of out-of-work Americans.

To pundits and the public, it seemed that the compromise
was shaping up to be a GOP victory: the Republicans would
preserve the Bush tax cuts for the rich and the Democrats would
obtain unemployment benefits already supported by a majority
in both the House and Senate (but blocked by a Republican filibuster). Republican hostage taking had paid off: they had forced
Obama to bend on a defining campaign promise.

At noon that day, Obama and his top advisers gathered again
in the Oval Office. McConnell had not provided Biden a firm
counteroffer. But he had signaled that the Republicans would
probably go for a payroll tax holiday. Refundability for the tax
credits, though, would be tough. And no extension of the green
tech tax credit.

McConnell had told the vice president he could not agree to
any of this without Lincoln-Kyl.

Obama and his aides pondered all the trade-offs. Did they
want to yield on Lincoln-Kyl? Or mount a fight and sacrifice
those other benefits?

Sperling was sickened by the thought of including Lincoln-
Kyl in the deal. But he calculated that a face-off over that would
blow up the whole package, and, more important, he was guided
by the numbers. He laid them out for the president and the others.

Lincoln-Kyl would provide about $20 billion to its benefi-
ciaries. The tax credits the White House was pushing for would
deliver between $50 and $70 billion in relief to middle- and low-
income Americans.

"It worked out to at least a two-to-one trade-off between tax
relief for our people and tax relief for dead rich people," one participant said.

Obama was still not willing to accede to a big estate tax break without refundability. He gave Biden his marching orders: get everything we want, and, by the way, don't give them everything they want.

Biden tried that. McConnell held firm: no refundability without Lincoln-Kyl. Come on, Biden said, isn't there middle ground on the estate tax? After all, some Democrats were already supporting a less generous estate tax proposal. McConnell agreed to discuss this notion with his fellow Republicans. But he didn't expect them to go for it.

That day, Pelosi called the White House: she wanted to bring a bigger group of Democratic leaders to see Obama to discuss the pending deal. White House officials persuaded Reid to bring his own leadership team. But White House aides decided to keep the groups separate as a way of preventing the opposition within their own party from reinforcing itself.

By that night—with Sperling and Jason Furman, the principal deputy director of the National Economic Council, negotiating details with GOP staff on the Hill—the White House had received word from McConnell that the Republicans would accept a payroll tax holiday and an extension of unemployment benefits in return for continuing all the Bush tax cuts for two years. But the Republican Senate leader still insisted on the Lincoln-Kyl estate tax provision.

THE NEXT MORNING, OBAMA, BEFORE DEPARTING FOR A SHORT trip to Winston-Salem to speak on education, met with Biden in the Oval Office.

The president said that accepting the estate tax provision on its own would be too much. The White House had to have more. Tell McConnell that the refundable tax credits were the price.

Obama was committed to wringing every last drop of stimulus he could out of the Republicans.

On Monday afternoon, Pelosi arrived at the White House with Democratic Representatives George Miller, Rosa DeLauro, John Larson, Xavier Becerra, Chris Van Hollen, and Steny Hoyer. They believed that not much had changed in the past two days, and the White House was merely informing them what items were currently on the negotiating table. They were not told how close the White House and McConnell were to a final handshake.

The meeting started in the Roosevelt Room without the president; the vice president walked the legislators through the various elements: payroll tax holiday, education tax credits, a child tax credit, refundability. He added that to get refundability, the White House had to concede on the estate tax.

Several of the Democrats recoiled. "We were very clear that the estate tax could be the straw that breaks the camel's back, and we would not support the package," one of the Democrats subsequently said.

While this was going on, Obama was in the Oval Office with his senior aides. When Phil Schiliro came in, the president asked for a report on the meeting.

"It's brutal," Schiliro said. "They're all worked up."

Obama left his office and walked the few steps to the Roosevelt Room. He sat down at the table. Obama gave no sign a deal was imminent.

The conversation got heated. Several lawmakers pointed out that this sort of compromise would violate Obama's campaign promise to kill the Bush tax cuts for the rich. Obama noted that he had also promised not to raise taxes on the middle class. A few urged Obama to lead a last-minute crusade against the Republicans for siding with the rich.

"They didn't realize that we were way beyond that point," a participant later recalled.

Can you take me step-by-step through your strategy? Obama asked the lawmakers who were advocating a different course. All tax bills, he reminded them, had to start in the House of Representatives, which in weeks would be in the hands of Boehner and the Republicans. So, he wondered aloud, how could legislation next year end up being any better than the package at hand?

"It was really clear they did not have a strategy," a senior administration official later said.

As Obama saw it, what these Democrats were requesting—three weeks of rhetorical debate during the holiday season—wouldn't change American politics. Worse, it would prevent any action on Don't Ask/Don't Tell, New START, and the DREAM Act.

As this meeting was under way, Democratic senators arrived at the White House for a similar session. They were escorted to the Cabinet Room, across the hall from the Roosevelt Room. For a while, Obama and Biden were going back and forth between the two meetings. The conversation with Reid and his colleagues was less stormy. Reid was no longer as ticked off. White House aides thought he seemed resigned.

While these meetings continued, Biden left to take a call from McConnell, who told the vice president they had a deal. Biden called the president out of one of the meetings and gave him the news and then returned to the Roosevelt Room. He didn't tell the House Democrats about the call. When the lawmakers left, most thought there would be more discussions.

Van Hollen was in a car heading back to Capitol Hill when he received an e-mail: the media were reporting that the White House and the Republicans had reached a compromise, and the Lincoln-Kyl estate tax cut was in it. Van Hollen called Pelosi.

"You're not going to believe this," he said.

Obama wasted no time. He shortly headed to the White House briefing room and announced that a deal had been hammered out.

THE PRESIDENT INSISTED THAT HE PREFERRED ONLY TO EXTEND tax cuts for the middle class, but told the reporters it would be "a grave injustice" and "a serious blow to our economic recovery" to allow tax rates to increase for these citizens. Republicans, he noted, were willing to block tax cuts for the middle class unless the rich were taken care of, but he could not let taxes be raised on the middle class—nor was he willing to permit two million out-of-work Americans to go without unemployment benefits.

Striving to position himself as a pragmatic leader, Obama equated Democratic partisans with Republican partisans, pointing out that "some people in my own party and in the other party . . . would rather prolong this battle." The message: I'm different from the rest of Washington, and I'm trying to make this town of political squabbling function.

He announced all the concessions he had forced out of the Republicans, but vowed to continue opposing the "tax cuts for the wealthiest Americans and the wealthiest estates."

His final point was a jab at those who had counseled battle: "These are not abstract fights." Millions of Americans facing tough times would lose unemployment insurance or see their taxes go up absent a deal. This package, he maintained, would fuel economic growth and job creation. He did not use the word *stimulus*.

OBAMA HAD ARGUABLY WON THE IMMEDIATE POLICY BATTLE with the Republicans. For yielding on the tax cuts for the rich, he had gained $238 billion in stimulus: $112 billion for the payroll tax holiday; $56 billion in unemployment insurance; $22 billion in equipment expensing; and $48 billion in refundable tax credits for child care, the earned income tax credit, renewable energy grants, and the education tax credit.

The Republicans had pocketed $91 billion in the top-bracket tax cuts and $23 billion for the estate tax measure.

This math was on Obama's side. But could the president now win the political and messaging war?

Minutes after Obama gave his statement, Gene Sperling, Jason Furman, and Dan Pfeiffer, the White House communications director, were on a conference call with reporters. They called this "an excellent deal" that included provisions that only a few weeks ago no one thought Obama could move past Republican opposition.

The Democratic-controlled House had recently failed to pass a bill to extend unemployment benefits for three months. Obama, they said, got thirteen months. The payroll tax holiday would mean $800 in relief for a worker earning $40,000. More than ten million low-income families, among them eighteen million children, would continue to receive a refundable child tax credit. Millions of students would receive the education tax credit included in the Recovery Act for another two years.

Yet the initial media coverage emphasized the extension of the Bush tax cuts—not Obama's maneuvers to obtain progressive policies benefiting tens of millions of Americans. The *New York Times* reported that the deal marked "a retreat from [Obama's] own positions and the principles of many liberals." *Retreat*—the *Times* had called it a retreat.

Congressional Democrats howled about the deal. Pelosi sent out a tweet that seemed to signal opposition: "GOP provisions in tax proposal help only wealthiest 3%, don't create jobs & add tens of billions to deficit."

Progressive activists accused Obama of betrayal. Bending on the Bush tax cuts for the rich *and* conceding on the estate tax was what mattered for them, not the other side of the deal. Nor did they realize that Obama was trying to clear the path for a wider agenda: Don't Ask/Don't Tell, New START, and more.

Other progressive policy wonks did spot the value in the deal. Robert Greenstein issued a statement noting that "the White House achieved everything it sought for low- and middle-income families." The Center for American Progress pointed out that the compromise would save or create 2.2 million jobs. Yet the positive reviews of these policy mavens did little to ameliorate Democratic anger.

The morning after the deal was unveiled, Obama met at the White House with several of the nation's leading progressive economists: Joseph Stiglitz, Alan Blinder, Paul Krugman, Robert Reich, Lawrence Mishel, and Jeffrey Sachs.

Obama wasn't expecting a hero's welcome from this crowd. Krugman, a high-profile columnist for the *New York Times* and Nobel Prize–winning economist, had recently excoriated him for even considering a compromise with the Republican "blackmailers." Krugman feared that any continuation of the Bush tax cuts for the rich would lead to a permanent extension—which would blow a $700 billion hole in the federal government's budget and push aside necessary investments.

Reich had recently written an article that slammed Obama for bolstering the Republican story of the economy ("big government, bureaucrats, and the cultural and intellectual elites" were responsible for the lousy economy) by setting up a deficit commission and freezing federal pay and discretionary spending. Obama, he complained, had failed to advance the competing plot line: Wall Street had brought the economy to its knees, and government action was necessary to undo the damage.

When Obama entered the room, he jokingly greeted the economists: "The last Keynesians in America."

Obama tried to assuage the economists about the tax-cut deal. Krugman was his "grumpy self," one participant said. The economists noted that the concentration of wealth and income at the top was a critical problem, and Obama acknowledged this.

But when Reich maintained that Obama was contributing to the GOP narrative by fixating on the deficit rather than jobs and by failing to blame the current economic woes on Wall Street malfeasance and overly pro-corporate policies, Obama disagreed. The American public, he said, was not ready to hear that, for it had been fed the government-is-too-big story for so long and people wouldn't listen to other arguments. Several of the economists noted that if Obama did not promote the counternarrative, the GOP message would dominate.

The meeting didn't change anything. "Given how he shook our hands at the end," one of the economists recalled, "it seemed he didn't want to see anyone in that room again."

In his next column, Krugman reported that he remained "deeply uneasy" about the deal. He accused Obama of having empowered hostage takers.

The day after the meeting, Reich wrote: "Apart from its extraordinary cost and regressive tilt, the tax deal negotiated between the President and the Republicans has another fatal flaw. It confirms the Republican worldview."

Reich was making an important point: Obama still didn't have a compelling economic tale to tell. His opponents, though, had a coherent take: government spending was out of control and harming the economy.

This was a misguided view. Government spending had not triggered the economic crash of 2008. Wall Streeters selling complex securities attached to lousy subprime mortgages had spurred a financial crisis that precipitated an economic collapse that had cost eight million people their jobs. Moreover, with the economy barely growing, many economists believed it was the wrong time to diminish demand by slashing government spending. But the Republican bumper sticker—government is the problem, not the answer—was simple and resonated with voters upset about the economy.

Obama was not countering that basic talking point. He didn't offer Americans a target for their fury or unease about the economy (while the GOPers did: the government). He didn't provide a conceptual road map for improving the economy in the short term (while the GOPers did: downsize government). His pitch for the tax-cut deal was that it would assist financially stressed American families—a noble cause.

But he couldn't describe it as a second stimulus because he had been outplayed by the Republicans on the first stimulus. Obama was doing what he could for the economy—using government action to gin up economic demand—but he wasn't fully explaining what he was achieving.

SOON AFTER THE UNSATISFYING SESSION WITH THE ECONOMISTS, Obama held a press conference to discuss the deal. In an opening statement, he hit what was for him the basic point: he had sidestepped a "protracted political fight" to help middle-class Americans.

And he had. But with the first question, Ben Feller of the

Associated Press challenged the president's ability to stand on principle: "You've been telling the American people all along that you oppose extending the tax cuts for the wealthier Americans. . . . So what I'm wondering is when you take a stand like you had, why should the American people believe that you're going to stick with it? Why should the American people believe that you're not going to flip-flop?"

This question reflected the dominant analysis: Obama had broken a prominent campaign vow for the sake of political expediency—not that he had sacrificed a stand to assist millions of Americans.

"Hold on a second, Ben," Obama shot back. "This isn't the politics of the moment. This has to do with what can we get done right now."

He said the Republican senators were not budging, and he had no choice; otherwise, the average American family would have to pay $3,000 more in taxes and two million Americans would be without unemployment insurance: "If there was not collateral damage, if this was just a matter of my politics or being able to persuade the American people to my side, then I would just stick to my guns."

Moments later Chuck Todd of NBC News asked whether Obama was telegraphing his negotiating strategy for coming fights with the Republicans: "They can just stick to their guns, stay united, be unwilling to budge . . . and force you to capitulate?"

In the media world, Obama had capitulated.

Obama didn't challenge Todd's use of that word. Instead he explained: "This is a very unique circumstance. This is a situation in which tens of millions of people would be directly damaged and immediately damaged, and at a time when the economy is just about to recover. . . . My point is I don't make judgments based on what the conventional wisdom is at any given time.

I make my judgments based on what I think is right for the country and for the American people right now."

Axelrod was sitting in the briefing room as Obama explained this point. The strategist was, as was often the case, staring at his BlackBerry and reading messages. But as he heard Obama say these words, he looked up and said to himself, *This is a* West Wing *moment.*

He and other White House aides were proud of Obama for placing policy and concern for struggling citizens ahead of politics and partisan demands.

In the final question, Jonathan Weisman of the *Wall Street Journal* noted that some on the left were questioning the deal and wondering if Obama possessed core values. What are you willing to "go to the mat on"? he asked.

Obama pledged he would fight the Republicans over making the Bush cuts for the rich permanent. He called that "a line in the sand." Obama then provided a version of his governing philosophy, presenting it as a counter to his critics on the left:

> This notion that somehow we are willing to compromise too much reminds me of the debate that we had during health care. This is the public option debate all over again. So I pass a signature piece of legislation where we finally get health care for all Americans, something that Democrats had been fighting for for a hundred years, but because there was a provision in there that they didn't get that would have affected maybe a couple of million people, even though we got health insurance for 30 million people and the potential for lower premiums for 100 million people, that somehow that was a sign of weakness and compromise.
>
> Now, if that's the standard by which we are measuring success or core principles, then let's face it, we

will never get anything done. People will have the satisfaction of having a purist position and no victories for the American people. . . . In the meantime, the American people are still seeing themselves not able to get health insurance because of preexisting conditions or not being able to pay their bills because their unemployment insurance ran out.

Obama was riled up—the public option controversy still stung—and he continued defending compromise as a necessary means of reaching long-term progressive goals:

This is a big, diverse country. Not everybody agrees with us. I know that shocks people. The *New York Times* editorial page does not permeate across all of America. Neither does the *Wall Street Journal* editorial page. . . . [P]eople have a lot of complicated positions, it means that in order to get stuff done, we're going to compromise. This is why FDR, when he started Social Security, it only affected widows and orphans. You did not qualify. And yet now it is something that really helps a lot of people. When Medicare was started, it was a small program. It grew. Under the criteria that you just set out, each of those were betrayals of some abstract ideal. This country was founded on compromise. I couldn't go through the front door at this country's founding. And if we were really thinking about ideal positions, we wouldn't have a union.

So my job is to make sure that we have a North Star out there. What is helping the American people live out their lives? What is giving them more opportunity? What is growing the economy? What is making us more competitive? And at any given juncture, there are going to be times where my preferred option, what I am absolutely positive is right, I can't get done. And so then my question is, does it make sense for me to tack a little bit this way or tack a little bit that way, because I'm keeping my eye on the long term and the long fight—not my day-to-day news cycle?

Obama was on a roll. He was not just defending the tax-cut deal. He was defending his entire presidency—not from the barbs of his rabid Obama-is-a-secret-Muslim-socialist foes on the right, but from the criticism of his purported allies on the left: "To my Democratic friends, what I'd suggest is, let's make sure that we understand this is a long game."

It looked as if Obama was more upset with the Left for not applauding this deal than he was with inflexible, filibustering Republicans for causing the dilemma. Obama believed he deserved more credit for this hard-fought compromise and for his previous accomplishments—and he wanted more backing from the Left for the tough decisions he'd have to make in the coming years while dealing with the recalcitrant GOPers.

Obama was right on this: the tax-cut deal had become a defining moment for his presidency.

OBAMA'S PRESS CONFERENCE PLEA FOR UNDERSTANDING AND backing did not persuade irate Democrats. At a meeting of the House Democratic caucus, Vice President Biden defended the deal and encountered angry tirades. Look, guys, Biden said, we did the best we could—and this would help the unemployed and the entire economy. (Noted forecaster Mark Zandi, who had advised McCain during the 2008 campaign, was predicting the provisions Obama had inserted into the package would increase economic growth in the coming years by 1.25 percentage points and create more than 1.5 million jobs.) But Biden didn't win over the caucus. "Not everyone is a pragmatist," a House Democratic leadership aide later explained. "Some people believe in rhetoric and the positive aspect of waging a good fight."

AXELROD AND SUMMERS JOINED GIBBS FOR THE DAILY PRESS briefing on December 8. Summers handled the economics. With the deal, he said, there would be economic growth; without, there would be the risk of a "double-dip" recession. This was a big chip to put on the table. But Summers had long believed that if taxes went up for middle-class families, the economy could receive a dramatic shock.

Summers was more circumspect on another matter. Is this "fiscal stimulus"? a reporter asked him. "Yes," Summers said, "it is putting money in people's pockets."

"So," the journalist followed up, "this is a second stimulus?"

Summers pulled back: "No, I don't think that's the right way to think about it at all."

"So this fiscal stimulus is a nonstimulus package?"

"I'm not going to play semantic games for you," Summers replied. "These are fiscal measures that promote growth."

The White House still couldn't call the policy what it was.

THE WHITE HOUSE HAD LITTLE PATIENCE FOR THE LEFT-OF-center nay-saying—seeing it as another sign that progressive activists and intellectuals were out of touch with reality and not truly representative of the base they claimed to speak for. This deal would help millions. Nothing close to it had been in play on Capitol Hill. And polling showed that the agreement played well with Obama supporters and liberals, as well as independents.

Toward the end of the week, conservative commentator

Charles Krauthammer published a column declaring that "Obama won the great tax-cut showdown of 2010—and House Democrats don't have a clue that he did." Krauthammer had seen through Obama's ploy: "If Obama had asked for a second stimulus directly, he would have been laughed out of town." Krauthammer was hardly pleased by this. He called the Obama-engineered compromise the "swindle of the year."

When Obama read the Krauthammer column, he was amused and also exasperated. He now found himself more in agreement with the conservative columnist than with the liberal columnist Frank Rich, who days earlier had derided Obama for becoming a victim of the Stockholm syndrome and for surrendering "his once-considerable abilities to act, decide, or think." Krauthammer got it, Obama thought, and Rich did not.

The tax-cut package passed without much trouble in the House and Senate (with close to half of the House Democrats rejecting it). When Obama signed the bill into law, both Pelosi and Reid skipped the signing ceremony. But McConnell was there.

That same day, Obama met at the White House with the top leaders of the labor movement. The unions had opposed the tax-cut deal. The unionists were also displeased with the South Korea trade accord Obama had recently signed and the administration's decision not to push for the Employee Free Choice Act, which would make it easier for unions to organize workplaces. Richard Trumka, the president of the AFL-CIO, complained, "We don't get real input" in White House decisions. He criticized Obama's recent decision to freeze pay for federal employees, noting that would prompt private sector employers to do the same. He suggested the White House was not doing enough to beat back the joint GOP–Chamber of Commerce attack on labor unions and collective bargaining across the country.

Obama defended the tax-cut deal as the only stimulus he could push through Congress and remarked that he was operating in a hostile political environment: "It used to be that you could govern by peeling off a couple of Republicans to do the right thing. Now Palin and Beck are the center of the Republican Party—and there is no possibility of cooperation."

Other union leaders vented their own frustrations. "You're too nice," Joe Hansen, the president of the United Food and Commercial Workers International Union, said to the president.

"Business leaders think all I do is bash them," Obama said.

"No one is more frustrated than me," the president told the group. "The Senate isn't working as a democracy anymore." He explained he had to make tactical concessions to enhance his political position: "You were nice not to kick me in the teeth over the pay freeze. I did it because the Republicans want to go further with layoffs and pay cuts. We can fight back because we put a pay freeze in effect. Giving up a little at the front end will help us fight a battle that's coming."

"We're losing white males," Obama pointed out, maintaining that was partly due to cultural issues: gun rights, gay rights, and race. "Fed by Fox News," he said, "they hear Obama is a Muslim 24/7, and it begins to seep in. . . . The Republicans have been at this for forty years. They have new resources, but the strategy is old."

This was Obama's plea: I'm doing what I can in difficult circumstances, I have to be crafty, I have to concede to proceed. Please understand, he asked.

Though Obama believed the tax-cut deal was a major victory, he felt he was stuck in a weak political position—up against a political culture decades in the making that denigrated the sort of government action the current crisis and the long-term needs of the nation demanded.

Rather than overpower this opposition, Obama was looking

for ways to outsmart it, to seize what he could of its agenda and rhetoric, to devise a combination of cunning negotiation and strategic (intermittent) confrontation that would land him on the higher political ground. He was planning ahead, anticipating a tough year of complicated calculations.

But planning ahead—and avoiding fights to better prepare for the next one—doesn't always work out as anticipated. In the frenzied days following the tax-cut deal, a moment suggesting the major trouble to come went largely unnoticed. It occurred during that press conference when Obama vividly defended the need to compromise. Toward the end, Marc Ambinder of the *National Journal* asked whether the outcome of the tax-cut talks would affect future negotiations with the Republicans on raising the debt ceiling.

Ambinder was peering into the future. In several months, the US government would have to lift the cap on the debt it could take on in order to cover the bills it had already accrued. Such a move would require an act of Congress. In years past, raising the debt ceiling had been a routine action supported by both parties because if the limit on borrowing wasn't increased, the US government could enter default, and that could spark a financial crisis. But in this current climate—when Tea Party Republicans were on an antispending crusade—could Obama depend on the GOP to abide by long-established norms?

Senior administration officials had considered securing an increase in the debt ceiling as part of the tax-cut compromise, but that would have meant yielding to further GOP demands—and such a move could have threatened the entire deal. So this had been abandoned.

Ambinder, though, presciently asked the president if the Republicans would hold the debt ceiling hostage and refuse to raise the cap unless the White House would agree to severe spending cuts.

Obama dismissed such a fear:

> I'll take John Boehner at his word that nobody, Democrat or Republican, is willing to see the full faith and credit of the United States government collapse, that that would not be a good thing to happen. And so I think that there will be significant discussions about the debt limit vote. That's something that nobody ever likes to vote on. But once John Boehner is sworn in as speaker, then he's going to have responsibilities to govern. You can't just stand on the sidelines and be a bomb thrower.

Moments earlier in this press conference, Obama had referred to Boehner as a hostage taker who didn't care about the governed. Yet now he appeared to be assuming that the hostage grabbing was done and that once Boehner and his colleagues were running the House, they would eschew their terrorist ways and behave reasonably.

"We were not Pollyannaish," Axelrod later insisted. "We knew the third year would become much more difficult."

But how much more difficult? Was Obama fully prepared for what was ahead? The president had done well in the recent negotiations, whether or not that was realized by the politerati or his fellow Democrats. Yet there were other high-stakes negotiations on the horizon—and other hostages to be seized.

Chapter Three

A COMPLETE WIN

JOE BIDEN WAS BEAMING. HE WAS STANDING AT THE PODIUM ON a stage in a crowded auditorium in the Department of the Interior.

"It's a real good day," the vice president exclaimed, and a crowd of several hundred people cheered.

In the audience were cabinet members, senators and representatives, and scores of gay rights activists. Three days before Christmas, they had come to witness a historic event—and the fulfillment of Obama's high-profile campaign promise to repeal the Pentagon's Don't Ask/Don't Tell policy banning out-in-the-open gays and lesbians from serving in the military.

On the stage with Biden were the two senators who had sponsored the legislation to end Don't Ask/Don't Tell: Joseph Lieberman, the Connecticut Democrat-turned-independent, much loathed by liberals for many reasons (including his hawkishness and his support for John McCain's 2008 presidential campaign), and Susan Collins, the moderate Republican from Maine. Next to them stood House Speaker Nancy Pelosi and Senate majority leader Harry Reid. Other honored guests on the stage included Admiral Mike Mullen, the chairman of the Joint Chiefs of Staff.

Though the mood in the room was celebratory, it was also a bittersweet event. That night, Congress would adjourn for the year. The Democrats' four-year-long control of the House was coming to an end; Pelosi was serving in her final hours as speaker.

When Obama walked on the stage, he gave Biden a quick hug. "Yes, we did!" the crowd chanted.

The president smiled at the assembled and said, "Yes, we did."

He told the story of Andy Lee, a gay GI, who during the Battle of the Bulge in World War II, scaled down the slope of an icy ravine—"with shells landing around him, amid smoke and chaos and the screams of wounded men"—to save a fellow soldier. The president quoted a special operations warfighter: "We have a gay guy in the unit. He's big, he's mean. He kills lots of bad guys. No one cared that he was gay."

Then the president strode to a desk on the stage, sat down, and signed his name to the law. Obama slapped his hand on the document, looked at the audience, and said, "This is done."

Signing the measure was one of the last acts of Obama's first two years in office. In two weeks' time, he would be facing a House controlled by Republicans beholden to the intransigent Tea Party wing of their party and a Senate with an enlarged Republican caucus. But for the moment, he and the others could exult in a pure victory—a success that had the hallmarks of an Obama accomplishment: patience, deliberation, determination, luck, tension, and nuanced behind-the-scenes deal cutting.

ABOUT A YEAR EARLIER, OBAMA CALLED JIM MESSINA, HIS deputy chief of staff, into the Oval Office.

"You're doing Don't Ask/Don't Tell," the president said.

"Why me?" Messina asked.

"Because I want to win."

By this point in the presidency, Messina had earned a nickname he fully embraced: Obama's fixer. The fortysomething rail-thin Capitol Hill veteran had been brought into the 2008 campaign by David Plouffe to serve as chief of staff, with the task of coordinating the massive political, field, and policy operations.

Prior to that, much of Messina's career had been devoted to Senator Max Baucus, as either chief of staff or campaign manager for the Montana centrist. Messina quickly became an integral member of Obama's inner circle. In the White House, Messina, an aggressive but low-profile fellow (who drove a Porsche), worked well with the hard-charging and profane chief of staff Rahm Emanuel. Messina once referred to Emanuel as the "smartest political strategist of his generation." And he readily acknowledged his own outsized ego.

"I personally couldn't tell you what he believes in," said one Obama campaign staffer who collaborated with Messina. "But he does reflect one distinguishing Obama quality: pragmatism."

In the White House, Messina became known as the guy who could take care of the president's problems. A cabinet nominee with a tax problem? Messina smoothed it over in the Senate. Tracking how various agencies would spend stimulus dollars? Messina was on it. He was the White House's equivalent of the Wolf, the Harvey Keitel character in *Pulp Fiction*. One Messina associate noted that he once explained his effectiveness this way: "If the White House deputy chief of staff calls you, you take the fucking phone call."

Messina was all over the health care bill—cajoling the Senate and handling the progressive outfits and unions that were pushing for reform that went beyond the pending legislation. He helped concoct the administration's controversial deal with the pharmaceutical industry: in return for the industry promoting health care reform and promising $80 billion in unspecified cost savings over ten years, the White House agreed to walk away from two prominent Obama campaign promises (using the purchasing power of Medicare to negotiate lower drug prices and facilitating the importation of cheaper drugs from Canada and Europe). Progressives decried the deal as a giveaway (or bribe) to Big Pharma; the White House saw it as a savvy move that neutralized a potential foe of health care reform.

Messina was also the White House's point man in regularly dealing with left-of-center constituency groups—liberal advocacy shops, unions, abortion rights groups, environmental organizations, and gay rights groups. Many Tuesday nights, Messina met with officials of these outfits at a Washington hotel, through what was called the Common Purpose Project, and discussed White House plans, priorities, and messages. Some of the outside participants considered the meetings merely sessions where the White House tossed out talking points and marching orders—which were ignored by the outside groups at their own peril. As some advocates saw it, if you didn't play ball the way Messina wanted, you'd be cut off.

OBAMA HAD RUN FOR PRESIDENT AS A CHAMPION OF GAY RIGHTS (with the notable exception of gay marriage), and in his first White House meeting with Gates, Mullen, and the chiefs of the military, the new president notified them that he intended to keep his campaign vow to end Don't Ask/Don't Tell.

The military men might have wondered if the fresh president was merely talking the talk. Sixteen years earlier, another new-to-the-office and young commander in chief with no prior military experience had kicked off his presidency contending that he would implement his campaign pledge to allow gays and lesbians to serve openly in the armed services. And it turned into a mess for Bill Clinton.

Not only did Clinton fail to lift this ban, but his bumbled efforts also prompted the Democratic-controlled Congress in 1993 to pass legislation enshrining Don't Ask/Don't Tell into law—prohibiting the military from questioning service members about their sexual orientation but allowing it to toss out soldiers who publicly indicated they were gay. More than twelve thousand members of the armed services would be forced out in the years that followed.

With this back story, Pentagon officials had cause to suspect Obama might not be eager to shove the Pentagon toward social change. But there was another reason for the military brass to believe that this campaign promise would be sidelined: the new president needed the military. He had two wars to wage. He was committed to withdrawing US troops from Iraq to end that war. He had pledged to revise the strategy guiding the US military effort in Afghanistan, hoping to achieve tangible progress in what increasingly seemed to be a morass.

Obama, a liberal president who had not spent much time during his short political career on military matters, had to establish a sound working relationship with the Pentagon in order to achieve his top national security priorities. Starting with an in-your-face initiative would likely not yield a productive partnership.

Before even moving into the White House, Obama had concluded that he could not win a fight over Don't Ask/Don't Tell without the Pentagon on his side. In the House and Senate, he would need the votes of centrist Democrats (and perhaps a few Republicans) who were close to the military. If the Pentagon were to openly (or privately) resist, the president figured, he would not be able to corral these votes.

As commander in chief, Obama could cut orders that would eviscerate Don't Ask/Don't Tell. He could instruct the Defense Department to halt or limit its enforcement. He could tell Justice Department attorneys not to defend the policy when challenged in court. "Many of his supporters believed that if he thought it was unfair, he should just get rid of it with a stroke of the pen," Gibbs recalled. But if Obama chose any of these avenues, it would be a declaration of war against the Pentagon. And any policy change he implemented on his own could be undone by a subsequent administration. Obama desired a complete— "durable," he often told aides—end to the ban. That meant he needed a law. That meant he needed Congress. And that meant he needed the Pentagon.

DURING THE FIRST YEAR, OBAMA DIDN'T CHARGE TOWARD repeal, but he was not letting the matter slide. As he ended one meeting with Gates and Mullen in the spring of 2009, he said, "You know I'm serious about repealing Don't Ask/Don't Tell."

He told them to start thinking about how to accomplish this. Jim Jones, a former Marine Corps general whom Obama had picked to be his national security adviser, was at this meeting. Afterward he cornered Greg Craig, Obama's White House counsel, in the hallway to ask if Obama was serious about forcing the Pentagon to drop the ban.

"Weren't you listening during the campaign?" Craig replied.

Not really, Jones said.

Throughout that year, Obama didn't lean on Gates and Mullen. Certain leaders of the gay rights community were not surprised. They were aware of the history (the Clinton disaster) and understood the dynamic (Obama had to win over the Pentagon before changing it). They were willing to be patient. Obama had an economy to rescue, wars to manage. Other gay rights advocates, however, were angry that Obama was not moving aggressively on their issues—not on Don't Ask/Don't Tell, not on financing AIDS programs, not on legislation to prohibit workplace discrimination on the basis of sexual orientation. (Some were still upset that the president had tapped Rick Warren, the popular evangelical pastor who had compared homosexuality to incest and pedophilia, to deliver the invocation at his inauguration.)

Toward the end of 2009, White House aides were thinking that the administration had to get moving soon on the repeal of Don't Ask/Don't Tell, if only to show that Obama would stand by his convictions—a matter that concerned Obama supporters beyond those focused on gay rights. After all, many did believe that as commander in chief Obama could reverse Don't Ask/Don't Tell with a snap of his fingers—an impression that led to the question: Well, why hasn't he? (Months earlier, the administration had enraged gay rights supporters when Justice Department lawyers, defending against a challenge to the 1996 Defense of Marriage Act, filed a legal brief noting that gay marriage was akin to incest and threatened traditional heterosexual marriage. Eventually, Obama would order the department to stop defending the constitutionality of DOMA.)

But the White House wasn't ready yet to move on this front. In November 2009, Joe Solmonese and David Smith, two officials of the Human Rights Campaign, a gay rights lobby, huddled with Messina. Their organization, they told him, was prepared to mount a major campaign to eliminate Don't Ask/Don't Tell. "I need you guys to hold off," Messina said, "until we finish the Afghanistan review." At the time, the president was conducting an intensive assessment of policy in Afghanistan. Once that was done, Messina told them, Obama would kick into gear on Don't Ask/Don't Tell.

Toward the start of 2010, it was time for Obama to forge a deal with Gates and Mullen. The White House assumed that Gates and Mullen were not eager to implement this big change— but not because of any strong personal feelings. When Gates had been CIA chief under the first President Bush, he had stopped the agency from investigating the sexual orientation of employees during security clearances. And Mullen, a sixty-three-year-old churchgoing Catholic, had grown up in liberal Hollywood, where his father had been an actor-turned-press-agent whose clients included Ann-Margret and Julie Andrews. He was raised in what was probably a tolerant environment.

But Gates and Mullen certainly weren't looking for a politically charged task in the middle of two wars. Supporting repeal would place them in opposition to some of their best friends on Capitol Hill—defense hawks who did not favor allowing gays and lesbians to serve openly. Representative Ike Skelton, a Missouri Democrat who chaired the House Armed Services Committee, was fervently against repeal. In the Senate, John McCain, the senior Republican on the Armed Services Committee, was poised to lead the opposition, even though he had once signaled he was open to ending the ban.

The two military men, though, told Obama that they would support repeal—but they had a condition: they wanted to be in charge of how it would happen. Gates demanded nine months or so for the Pentagon to study how to handle a transition—and he requested no White House action on repeal until after the study was completed. That was the price for their cooperation.

Obama consented. Gates and Mullen would accept his policy of change, and he would let them control how the change occurred. The president had enlisted them as partners in this mission. In his State of the Union address on January 27, 2010, Obama declared he would "work with Congress and our military" in the coming year to end Don't Ask/Don't Tell.

Obama had traded time for the support he deemed essential. But Democrats on the Hill were not so patient. Several Democratic senators and representatives were already planning on pursuing an amendment to the annual defense authorization bill—which sets overarching guidelines for military programs— that would eliminate the ban. Despite Obama's agreement with Gates and Mullen, the White House was in danger of losing control of the process.

The day before Gates and Mullen were due to testify before Congress on Don't Ask/Don't Tell, Messina convened in the White House a meeting of Washington's leading gay rights advocates, including Solmonese and Smith. This is what's going to happen, Messina told the advocates: Gates and Mullen will testify that they support repeal but that they don't want legislation yet—not until they have had time to come up with a plan of their own. And it will take them about nine months to conduct a study—not on *whether* to repeal, but on *how* to repeal.

This is great, the advocates said. But they pointed out that if the White House were to wait that long for the Pentagon to finish its study, there would be no vote on repeal legislation until *after* the November congressional elections—and that was rather risky, given the widespread belief that the Republicans would win the House. Besides, the advocates added, legislation was already moving on the Hill, and the grassroots were fired up and ready to roll. Why wait?

The timing was not open to discussion, Messina said. He and the other White House aides sat there stone-faced, as it dawned on the advocates they were not being asked for their opinion; they were being told what had already been decided. This was not full consultation; Obama and his aides had already devised their game plan. This was notification.

Messina's message was not a shock for the advocates. Since the start of the administration, policy pushers and issue lobbyists had complained that the Obama crowd was not as collaborative as they had hoped. During the 2008 campaign,

Obama and his advisers had routinely described their endeavor as a grassroots movement with ownership that extended far beyond the candidate and his political crew. That was true in a macro sense; never before had so many millions of voters plugged into a campaign through volunteering, donating, and social-media networking.

Yet when it came to operational control, David Plouffe had held the reins tight. More notable, he, Axelrod, and Obama had often eschewed—that is, fought off—advice from outside their small circle. "No organization can survive flitting from thing to thing, trying to please outside observers," Plouffe wrote after the campaign.

Still, Obama and his advisers had believed the 2008 campaign belonged to their millions of supporters. They did not appear to feel the same about his presidency. The pressures of governing—particularly at a time of multiple crises—did inhibit authentic and extensive outreach that could be incorporated into decision making. But Obama and his aides had essentially placed his grassroots campaign army on hold, with the Obama for America volunteer powerhouse folded into the Democratic Party and renamed Organizing for America. Obama and his small group of advisers were the deciders. The president would mainly be working *for* his supporters, not *with* them.

THE GAY RIGHTS ADVOCATES LEFT THE WHITE HOUSE WITH mixed feelings. It was a tremendous boost to have Gates and Mullen endorse repeal. But they feared Obama's strategy—which could stretch out the process for a year and a half—would fail. And senators and representatives would be galloping ahead with repeal legislation that had a shot at success. Obama's go-slow timetable seemed a formula for conflict and squabbling among all the players supporting repeal.

GATES AND MULLEN DID HOLD UP THEIR END OF THE BARGAIN. In a much-covered appearance before the Senate Armed Services Committee on February 2, 2010, they testified in favor of repeal.

It was a heartfelt moment for the military's top officer. Mullen declared that "allowing gays and lesbians to serve openly would be the right thing to do." He added, "I cannot escape being troubled by the fact that we have in place a policy that forces young men and women to lie about who they are in order to defend their fellow citizens."

Gates appeared a bit less enthusiastic. "We have received our orders from the commander in chief, and we are moving out accordingly," he testified. The defense secretary noted that it would take the Pentagon about a year to study how to implement a new policy.

This was a historic hearing—the top brass backing Obama's call for repeal. But John McCain was not enjoying it.

In 2006, he had remarked on MSNBC's *Hardball,* "The day that the leadership of the military comes to me and says, 'Senator, we ought to change the policy,' then I think we ought to consider seriously changing it." Now he was steaming mad and ready to lead the opposition. (Though McCain's need to survive a tough Republican primary battle in Arizona might have motivated his new hostility toward repeal, administration officials and Washington pundits couldn't help but wonder if he was acting out his lingering resentment toward the man who kept him from the Oval Office.)

And not all Democrats were eager to dump Don't Ask/Don't Tell. Representative Steny Hoyer publicly said that he wanted Democrats in their home districts to focus on "jobs and fiscal responsibility," not gays in the military. In private conversations with White House aides, Hoyer complained about the president forcing House Democrats to take what he thought would be a difficult vote.

THOUGH GATES SEEMED TO BE ON BOARD, WHITE HOUSE AIDES worried about him. He had served in Washington many years and certainly knew how to sabotage the administration's repeal effort if he chose to do so. They wondered why the Pentagon needed so long to conduct the study. What would happen if the vote on the repeal were postponed to the next Congress? Messina had to remind his colleagues that if the Democrats lost the House and/or the Senate, there would be a lame-duck session. That could be their last chance.

"The gay rights community didn't trust the Pentagon on all this," a White House aide subsequently said. "They thought we were crazy."

Gates was adamant that no legislation go forward until the Pentagon finished its study. In April, he and Mullen sent a letter to Ike Skelton, chairman of the House Armed Services Committee, warning Congress not to undo the ban until after the Pentagon study was completed—and the due date was December 1. That caused more worry for gay rights activists. After that date, there likely wouldn't be enough time to consider last-minute repeal legislation. They wondered if the Pentagon was trying to kill repeal with this schedule. And if this whole matter got kicked into the next year—and the Republicans were then controlling the House—repeal would likely be dead.

As a bill that would end the ban (after the president and the Pentagon certified that the armed forces would not be disrupted) approached a vote that spring, Obama did not take a public position on it. Some gay activists thought the president, who had vowed to be a gay rights champion, was wimping out at a crucial time. Yet behind the scenes, Messina and the White House were urging Democrats to support the measure. It was a delicate juggling act for the White House: support repeal without openly championing the bill (to avoid irritating Gates), but on the sly advising the bill's backers.

Messina had recruited Joseph Lieberman to be the main cosponsor of this amendment (with Susan Collins). Lieberman had long supported ending the ban on openly gay and lesbian service members. But he was not eager to be the point man without assurances that the White House was serious. He was not interested in a lonely crusade, and there was plenty of cause for Lieberman and the president to be wary of each other. Could Obama trust Lieberman to go to the mat, even if it meant fighting his dear friends McCain and Senator Lindsey Graham? Could Lieberman trust the president to have his back all the way? Messina persuaded Lieberman of the president's support, and the independent scorned by Democrats charged ahead.

Meanwhile, McCain was waging a war on repeal, circulating letters from the chiefs of the navy, army, air force, and marines that called for Congress to hold off on any legislation before the study came out. (Earlier in the year, Marine Corps Commandant General James Conway, who opposed repeal, had declared, "I would not ask our Marines to live with someone who is homosexual if we can possibly avoid it.")

Even so, the measure passed the Senate Armed Services Committee at the end of May on a 16–12, mostly party-line vote. Hours later, the House approved its version of the measure on a 234–194 vote. "We wanted to tell people what we were doing to pass the bill," one White House aide said. "We wanted to take credit. But we had to play the cards the right way."

Then momentum stalled. The Senate bill containing the repeal provision bogged down due to other issues, and in September, Reid and the Senate Democrats couldn't get the defense bill past yet another GOP filibuster.

White House aides fretted about running out of time on Capitol Hill. On several occasions, Pentagon officials mentioned to White House aides that the Defense Department might not meet the December 1 deadline for the report. "That's when we became unglued," a senior White House official later said.

BUT IN SEPTEMBER, OBAMA CAUGHT A LUCKY BREAK. A FEDERAL district judge in California ruled that the Don't Ask/Don't Tell policy was unconstitutional in a case that had been initiated by Log Cabin Republicans, a group of GOPers who backed gay rights. The judge rejected the Justice Department's request to temporarily suspend her order, pending an appeal. And the Pentagon had to rush to create guidelines for ending the policy in the jurisdiction covered by the order—before a higher court issued a stay. All this raised the specter of chaos for the Pentagon: a court could force it to end Don't Ask/Don't Tell at a moment's notice. Gates feared the worst—losing control—and this made him a fan of enacting legislation, and doing so before the Republicans gained control of the House.

Soon after the court ruling, Obama called the chiefs to the Oval Office and told them, If we don't repeal Don't Ask/Don't Tell in an organized manner with Pentagon input, the courts will eventually force you to end the policy without much, or any, warning. Army chief of staff General George Casey Jr. replied that the courts couldn't do that on their own.

The president was surprised that Casey wasn't acknowledging this very real possibility. Yes, they can, he responded, and they just might.

Obama didn't win over the chiefs. Moreover, he couldn't tell if the chiefs didn't fully understand the court system or were in denial. Soon after, White House aides were told that Gates, too, had met with the Joint Chiefs of Staff and informed them that it was necessary for the Pentagon to lift the ban on its own terms before the courts compelled the military to ditch Don't Ask/Don't Tell.

DURING A BUSY TIME—THE CONGRESSIONAL MIDTERM CAMpaign was in full gear, the White House was trying to figure out what could be done to improve the flagging economy, the Bush-tax-cuts battle was brewing—Obama kept a close watch on Messina. This was a top priority for the president, but he had chosen to win this change through a careful and deliberate process that was not in public view.

At an October town hall meeting, a Howard University professor grilled Obama on his failure to repeal Don't Ask/Don't Tell through executive order, as President Harry Truman had done in racially integrating the military in 1948. Obama insisted that he could not end the policy with a stroke of a pen, but said he had been able "for the first time" to get the chairman of the Joint Chiefs of Staff and the defense secretary to support repeal. He added, repeal "has to be done in a way that is orderly." It was classic Obama: attempting to achieve change through consensus, not confrontation.

"This policy will end, and it will end on my watch," he declared.

Orderly change was not a rousing rallying cry. For months, some gay rights activists had been screaming at and about Messina, fearful that he was not orchestrating a vigorous enough effort. Messina's response mirrored Obama's attitude: You got to stick to the game plan.

ONCE THE REPUBLICANS WON CONTROL OF THE HOUSE IN November, repeal was on the line. Time was running out. The defense bill in Congress was caught in the usual partisan gridlock. And White House aides had picked up rumors that the Pentagon study might recommend separate quarters or bathrooms for gay and straight soldiers. Anything of that sort would generate instant controversy and could blow up their best-laid plans. If anyone at the Defense Department wanted to double-cross the White House, placing over-the-top recommendations in the report would be an easy and clever way to do so.

But during a briefing in the Situation Room, Pentagon officials gave Messina, Schiliro, and Denis McDonough, the deputy national security adviser, a reassuring preview of the study. A Pentagon poll had found that many soldiers, especially younger ones, didn't care about serving with gays and lesbians. Obama's aides breathed a collective sigh of relief.

But this didn't guarantee action on Capitol Hill. In late November, the White House and Harry Reid's office were each unsure of the other's commitment to accomplishing repeal in the few remaining weeks. To Reid and his staffers, it seemed that the White House cared more about ratifying the pending New START treaty. "They would love both, but their first, second, and third goal was START," a senior Democratic Senate aide recalled.

But as White House aides saw it, Reid was constantly warning the White House that there was not enough time to take care of both New START and Don't Ask/Don't Tell, especially when a tax-cut deal was not yet worked out. (Months later, Democratic Senate staffers would note that Reid had heroically pushed for repeal and done the real heavy lifting; White House aides would contend that they, acting at the president's initiative, had soldiered on, despite Reid's pessimism. In Washington, such conflicting perspectives are not uncommon.)

ON NOVEMBER 30, THE PENTAGON RELEASED THE REPORT. IT was exactly what Obama had hoped for: repeal, it stated, posed no serious risks. Seventy percent of service members, according to the study, said that repeal would yield positive, mixed, or no consequences. In other words, America's fighters were not worried. At a press conference, Gates pointed out that ending this ban "would not be the wrenching, traumatic change that many have feared and predicted." He insisted it was a "matter of urgency" that Congress pass repeal before the lame-duck session ended.

There was still resistance from the brass. General Casey and Marine Corps Commandant General James Amos, testifying on Capitol Hill, both beseeched Congress not to alter the policy. Repeal, Casey warned, "would be a major cultural and policy change in the middle of the war."

Seizing upon this testimony, McCain threatened to filibuster the pending defense authorization bill with the repeal provision. Sixty senators—the number needed to break a filibuster—were now on record as supporting repeal, but Republicans had also pledged to stand with their party in opposing the defense authorization legislation until the Senate passed a tax-cut measure and Reid permitted more debate on amendments to the defense bill. (Reid contended there wasn't sufficient time to do that.) This meant that McCain likely had the forty-one votes for a filibuster.

On December 9, Reid tried to move the defense bill toward a vote, and McCain made good on his threat: the Republicans blocked the measure. News reports noted that this could well be the end of the repeal effort. It looked as if the Don't Ask/ Don't Tell ban might survive the lame-duck session. Solmonese accused both parties of engaging in "shameful schoolyard spats."

About this time, Brian Bond, a White House aide working on the issue, called leading gay rights activists to the White House. Some were worried the whole endeavor was on the verge of collapse and doubted Obama's commitment to the cause. Bond wanted to keep the activists galvanized. He was in a tough position. As a former leader of a gay rights group, he was often the target of the community's frustration—and often on the receiving end of a familiar complaint: Why isn't the president doing more?

Bond had earlier arranged for his boss, Valerie Jarrett, to address the dozen or so advocates. The plan was simple: she would say the president, *her good friend,* remained committed to repealing Don't Ask/Don't Tell. That presumably would have more weight than another yeah-we're-on-it from Bond.

As Jarrett was about to head to the meeting, she mentioned it to Obama. Bring them here, he said: If you're backing up Brian, why don't I just talk to them to back you up?

Bond escorted the group to the Roosevelt Room, and once they were seated, Obama entered.

"Something tells me you don't believe I'm committed to this," he told the group. "So rather than listen to Valerie or Brian, you can hear it from me."

Obama acknowledged that repeal seemed to be taking longer than it should. "I just want you to know," he said, "I'm going to get this done."

How? one of the activists asked.

Obama already had a Plan B in mind: a stand-alone bill to repeal Don't Ask/Don't Tell. That would take some doing, given the procedural obstacles to quick action on anything in the Senate. And Obama wasn't yet ready to discuss the possibility.

"You have to trust me," he said.

THE DAY AFTER MCCAIN LED THE FILIBUSTER AGAINST THE DEfense bill, Lieberman and Collins introduced that stand-alone bill to repeal the ban. The *Washington Post* called it a "Hail Mary." McCain was furious with his friend Lieberman.

It would be difficult to ram a new measure through the Senate, but Reid vowed he would use his power as majority leader to skip the committee process and bring the bill up for a quick vote. Pelosi declared that she could move an identical measure quickly through the House. And Obama's underutilized Organizing for America e-mailed its millions of supporters, asking them to pressure Republican senators. Reid would need a handful of GOPers to skirt a Republican filibuster attempt.

Robert Gates signaled that he was for quick congressional action: with no legislation, he said, "my greatest worry will be that then we are at the mercy of the courts and all the lack of predictability that that entails."

MUCH WAS HAPPENING ON CAPITOL HILL. THE FINAL DETAILS of the tax-cut deal were being hammered out. Vice President Biden, looking to round up Senate Republicans for ratification of the New START treaty, was in intense discussions with hesitant GOPers. During those talks, McCain and Lindsey Graham hinted at a deal: forget about repeal of Don't Ask/Don't Tell, and we could support the treaty.

They did manage to put it more subtly. They claimed that forcing a vote on repeal would create a sour mood among Republican senators, and opposition to New START would solidify. They were not offering a quid pro quo, but it was close. Their message was akin to a *Sopranos*-style warning: nice little arms control treaty you have here, Mr. Vice President; it'd be a pity if anything were to happen to it.

Hill Democrats heard rumors about this tempting offer and worried that the White House would not resist. Biden did view

McCain and Graham's implied threat as serious. His own aides could not tell if Biden thought it best to call the McCain-Graham bluff or cave to their threat. "Biden is the ultimate senator," recalled a White House official. "He was very concerned about our ability to get START ratified without making a deal. We were playing high-stakes poker. It was a nervous game for everyone."

The vice president conveyed the offer to his boss. Obama's response demonstrated his dry sense of humor. "Joe," he said, "I know you'll get the votes on START. You're just trying to make it easier on yourself."

Obama told the vice president he was not yielding on Don't Ask/Don't Tell, and he instructed Biden to pass that on to McCain and Graham. The president understood the risks: the ratification of New START now might be jeopardized.

ON DECEMBER 15, THE STILL-DEMOCRATIC HOUSE VOTED 250–175 for the stand-alone repeal bill. Fifteen Republicans backed the bill, fifteen Democrats did not. But this was the easy vote. Boehner did not whip the Republican conference against the measure. The real test was in the Senate.

For these final days, Messina practically moved into Biden's office in the Senate and closely tracked which senators were for or against the measure.

During the floor debate, an irritated McCain declared, "I hope that when we pass this legislation that we will understand that we are doing great damage."

A filibuster attempt flopped, with McCain coming up eight votes shy. Lieberman and Collins's repeal bill then sailed through on a 65–31 vote, with eight Republicans joining the Democrats.

Messina was standing off the Senate floor, with Joe Solmonese and other gay rights advocates—some of whom had often clashed with Messina over strategy. Valerie Jarrett was with them. As the measure passed, Messina and Solmonese were both in tears.

That same day, Obama fell short on another last-minute goal. The DREAM Act, which would provide a path to citizenship for undocumented student immigrants who had come to the United States as children, was killed by a GOP filibuster. With fifty-five senators voting for the Obama-endorsed measure, the bill was five below the supermajority needed to bring the measure to a final vote. Most Senate Democrats supported the measure, which had passed in the House, but five—Ben Nelson of Nebraska, Jon Tester of Montana, Mark Pryor of Arkansas, Max Baucus of Montana, and Kay Hagan of North Carolina— voted against it. And Democratic Senator Joe Manchin of West Virginia had ducked out of votes that day for a Christmas party.

Obama had come close; had he been able to persuade his own party comrades, he would have bagged another major lame-duck success. But he had at least obtained a political victory. With both parties needing to compete for votes in the growing Latino American community, Obama and the Democrats had placed the Republicans on record as thwarting legislation important for this crucial bloc.

For the president and his aides, repealing Don't Ask/ Don't Tell was a reaffirmation of his pragmatic approach. He had coaxed Gates and Mullen—even when doing so irritated his core supporters and alarmed gay rights advocates. He had not thumped his chest and demanded an end to a discriminatory policy. He had not relied on movement politics or called for sustained outside pressure. His nonconfrontational approach had yielded the results he desired. "He always felt that he could only do this if he had the Pentagon involved and that he had to walk them through the process," a White House aide observed.

There was another takeaway for Obama and his White House team: they prevailed because they stayed with the strategy. After the vote, Messina would routinely point out that the gay rights community had screamed at him for six months—while the president waited for the Pentagon report and time was running out—but that everything worked out because "you got to stick to the game plan."

Obama had benefited from at least one lucky bounce: the September federal court decision. The president's initial approach had left little margin for error or delay, and as 2010 went by, it did seem that granting the Pentagon so much time to produce the report could undercut the repeal effort. Yet that court ruling—which had resulted from the endeavors of Republican gay activists—had concentrated Gates's mind and prompted him to meet the same end-of-year deadline the president faced.

Coming two weeks after the tax-cut deal had been sealed, the repeal of Don't Ask/Don't Tell was a deliverable for die-hard Democratic progressives who had been upset by Obama's failure to stop the Bush tax cuts for the rich.

"Our base really fucking cared about DADT," one White House aide said. "A big amount." Obama cared, too. "He knew a lot of people thought it would be easy to do this," Gibbs later said, "when he had thought it would be hard."

Days after the Senate victory, Obama asked Messina to join him for the ride to the Interior Department building for the signing ceremony.

"This reminds our people that we can still do big things," the president told his deputy chief of staff. And, Obama added, it was also a reminder for the White House staff.

Many of Obama's successes had been either mixed (the tax-cut deal) or ugly (the health care initiative). This win was unadulterated and complete. "That day was a high point in the presidency for him," a senior administration official recalled, "in part because people had doubted him."

Chapter Four

A Quiet Victory

THE DAY AFTER THE SENATE VOTED TO REPEAL DON'T ASK/
Don't Tell, President Obama had to decide whether or not to risk
a Senate vote on ratifying the New START nuclear arms reduction treaty he had negotiated with Russia.

It was a chancy move because he couldn't be certain he'd
prevail. If the vote went badly, his standing at home would be
diminished, his concerted effort to bolster ties with Russia would
be in tatters, and his attempt to nudge the world toward zero
nuclear weapons would be discredited. He'd be seen as weak. No
vote would be better than a losing vote.

Given the Senate's starchy rules—and the few days left on
the legislative calendar—this day, December 19, was essentially
the deadline for Harry Reid to file a cloture petition that would
set up a vote to end debate on the ratification legislation and
allow the Senate to move toward an actual vote on the treaty.
Once the petition was filed, the cloture vote would have to occur
on December 21.

The White House was sure it had the sixty votes to overcome a Republican filibuster, yet if it succeeded on cloture, the

administration would essentially be locked into a vote on ratification. This was the dilemma: approving the treaty required sixty-seven votes, and Obama did not have that number of firm commitments. It was entirely possible he would win the cloture vote but fall short on the final vote. Obama conferred with Vice President Biden, who was in charge of ushering the treaty through the Senate. Biden believed the votes were there, even though he hadn't pocketed enough pledges.

The president didn't have the luxury of waiting until Biden nailed down the nine GOP votes the White House needed to win ratification. Obama had to decide now whether to pull the trigger, even if he might be committing himself to a defeat.

Let's go for it, he told Biden.

Biden felt confident. But not everyone in the White House trusted his intuition. That afternoon, Tom Donilon, the national security adviser (who had succeeded Jim Jones in this position two months earlier), organized a conference call for a handful of aides working on the treaty. Donilon had long worried that a losing vote would be an enormous strategic failure and was wary of any action that could lead to rejection.

On the call, Donilon raised the question: Should the White House proceed with the Senate vote? Louisa Terrell, a former Biden aide who worked in the White House legislative affairs shop, noted that they still didn't have a sufficient number of hard commitments from Republican senators.

Then Biden came on the line. What's this call about? the vice president asked.

Donilon said he was checking that the White House had the votes before the cloture petition was filed.

"Tom, we have the votes," said Biden, who had spoken to a dozen senators in recent days. Stop worrying, he added.

Donilon wanted to go through the Republican senators one by one. Biden wasn't interested. "I've personally talked to the senators," he said.

Donilon suggested having one more conversation with the president. Biden insisted the votes were there.

"I'm calling Harry. I'm telling him to file," Biden said.

The clock was now ticking.

IN APRIL 2010, OBAMA AND RUSSIAN PRESIDENT DMITRY MEDvedev signed the New START treaty, a follow-up accord to an earlier START treaty first proposed by President Ronald Reagan. (START stands for Strategic Arms Reduction Treaty.) The product of a year of US-Russian negotiations—which began soon after Obama took office—New START would compel both superpowers to reduce their strategic nuclear warheads to 1,550 within several years. (Russia maintained a stockpile of about 2,800 warheads and bombs; the United States, about 2,200.) It would also revive mutual monitoring of each nation's nuclear arsenal, which had ended in December 2009 when an earlier arms control treaty expired.

Though the treaty would lessen the number of nuclear missiles pointing at Americans, the agreement was a modest accord, certainly nothing radical, just a continuation of a process Reagan initiated. Yet it had significance beyond the warheads it would limit. The agreement was a component of Obama's attempt to improve relations with Russia, and the administration was hoping this effort would lead to more Russian pressure on Iran regarding Tehran's nuclear program.

There was another ancillary benefit. Obama, since the start of his presidency, had repeatedly stated his intention to lead the world toward a nuclear-free day. Administration officials and arms control experts considered approval of the New START treaty necessary to convince other countries—including aspiring nuclear nations—that Obama was serious about the march to zero nuclear weapons.

Before the ink was dry on the accord, conservatives and Republicans began complaining it would impede deployment of the missile defense systems they cherished. (There was nothing in the treaty that specifically set such restraints.) They moaned that the treaty had nothing to do with the more immediate threat of Iran's nuclear program. For some foes, the main objection seemed to be that this was Obama's treaty.

Senator Jon Kyl, the Arizona Republican and minority whip, was chief of the opposition. For years, he had been a stalwart foe of arms control treaties and a devoted fan of missile defense. He was allergic to the very idea of accords that limit nuclear weapons, huffing in a 2000 speech, "Honorable nations do not need treaty limits to do the right thing."

Kyl had long been grousing about one issue in particular: the modernization and maintenance of the existing US nuclear arsenal. He claimed that the federal government was not spending enough money on the upkeep of warheads. This debatable point was a separate issue from the number of world-destroying warheads maintained by the United States and Russia, but Kyl saw an opportunity to squeeze more funds for his pet cause. His fellow Senate Republicans viewed Kyl as their point man on arms control issues; thus, he controlled a bloc of votes. The White House would have to do business with him. If Obama couldn't win over Kyl, he'd have to neutralize him.

The administration began romancing Kyl before the accord was signed. In a February 2010 speech on nuclear security at the National Defense University, Vice President Biden noted that the president agreed that "our nuclear complex and experts were neglected and underfunded." He pointed out that the president had called for $7 billion for "maintaining our nuclear stockpile and modernizing our nuclear infrastructure." This was $624 million more than Congress had approved the previous year—and an increase of $5 billion over the next five years.

But Kyl wouldn't be bought off so easily. He, John McCain, and Joe Lieberman sent the White House a letter warning that they would oppose the treaty if the Russians were to claim the right to withdraw from the agreement in the event any US missile defense deployments altered the strategic balance. It was common for parties to an arms control agreement to reserve the right to pull out unilaterally. By making this a big deal, Kyl and the others were signaling that they would use whatever ammo they could dredge up to torpedo the treaty.

ONCE OBAMA OFFICIALLY SENT THE TREATY TO THE SENATE IN May for ratification, its fate was in the hands of Senator John Kerry, the Massachusetts Democrat who chaired the Senate Foreign Relations Committee. Like Biden, Kerry had long labored on arms control matters. Winning this battle was important to Kerry, who had been a key ally of the Obama White House on foreign policy matters. It was widely assumed within Washington that Kerry yearned to be secretary of state, should Hillary Clinton vacate the post. His chances of attaining that position could depend on how he managed this task.

Kerry and his staff immediately began collaborating with Biden and White House aides, especially Brian McKeon, Biden's deputy national security adviser who had for years worked on treaties in the Senate as a Biden aide. Their goal was basic: figure out how to navigate the Republican opposition.

Though previous arms control treaties had passed through the Senate with overwhelming support and little fuss, Kerry knew this time it would be different. The political climate was toxic.

Collaborating closely with Senator Richard Lugar of Indiana, the senior Republican on the Senate Foreign Relations Committee, Kerry developed a plan to gain Republican support. His secret weapon: keeping the partisan rhetoric low—an approach in sync with the institutional temperament of the White House. Obama and Kerry agreed that this was a fight to be won with finesse, not fisticuffs.

KERRY AND LUGAR HELD TWELVE HEARINGS ON THE TREATY, which featured officials from the Nixon, Ford, Carter, Reagan, Clinton, and both Bush administrations who testified in support of the agreement. The committee collected endorsements from seven former commanders of the US Strategic Command and past secretaries of state, including Republicans James Baker, Colin Powell, Condoleezza Rice, and Henry Kissinger.

Kerry and Lugar dutifully conveyed dozens of questions regarding the treaty from Kyl and other Republican senators to the White House for quick and thorough answers. Kerry's aim was to do everything possible to prevent ratification from becoming another one of those Washington mud-wrestling matches. Kerry would afford critics the chance to express their opposition. He would not hurry ratification—to deny Kyl and others a line of attack: *We can't vote for a treaty that is being bum-rushed through the Senate!*

No drama—that was the strategy.

THE FIRST MAJOR HURDLE WAS OBTAINING A STAMP OF APPROVAL for the treaty from the Foreign Relations Committee. The White

House and Kerry knew they could easily win the panel's okay on a party-line vote (with Lugar joining the Democrats). But that would create a D-versus-R story line. Instead, Kerry and White House staffers targeted two Republican senators—Tennessee's Bob Corker and Georgia's Johnny Isakson—as possible pickoffs. If they gained their votes, ratification would prevail on a 14–5 vote, an undeniable signal of bipartisanship.

While Kerry and Lugar courted Corker and Isakson, the White House lay low. And Kerry took the clever step of allowing Lugar to draft the resolution of ratification. This could make it easier for Republicans to support the bill. Kerry even griped publicly that he wasn't fully satisfied with Lugar's resolution—a way to make it more attractive to Republicans.

As is so often the case in Washington, timing was key. The hearings finished in mid-July. If the committee vote didn't happen right away, it could not occur until after the August recess—which would allow scant time for the full Senate to consider the treaty before Congress scurried out of town for the midterm elections. Obama wanted to wrap this up before the elections; otherwise, Republicans would inevitably argue that ratification ought to be postponed until the next (more Republican) Congress.

But Corker and Isakson were not yet in the bag. They said they needed more information from the administration on nuclear modernization. Corker was sympathetic to Kyl's concerns about funding for the nuclear stockpile; the Y-12 nuclear weapons facility was based in his home state of Tennessee.

At one point, Kerry and Lugar were at the White House discussing with Biden how to handle this important scheduling matter. Obama dropped by and listened as Kerry explained that a prerecess vote could create a partisan environment. Some White House aides were concerned that Kerry was following in the footsteps of Senator Max Baucus, the Democratic chairman of the Finance Committee, who, during the health care debate, kept pleading for more time to win over moderate Republicans. Yet he failed to do so. Was Kerry falling into a similar trap?

But Kerry believed he had little choice and postponed the committee vote. A ratification vote in the full Senate before the election was now out of reach.

Meanwhile, Kyl was pressing for more money and more information. "It looked as if Kyl intended to slow-walk the treaty to kill it," a Democratic Senate aide recalled.

As the mid-September committee vote approached, Obama did not get directly involved; there was no need. Lugar had extracted commitments from Corker and Isakson, and Biden called them to make sure they had no last-minute concerns. The committee approved ratification on a resounding 14–4 vote (Republican Senator Jim DeMint did not vote), but now the treaty was heading for the lame-duck session, where nothing was assured.

AFTER THE ELECTION, SOME OBAMA AIDES—MOST NOTABLY, Biden—were eager to forge ahead; others feared the Republicans would thwart a vote or, worse, defeat the treaty. National security adviser Tom Donilon considered any scenario that could end with a loss as perilous. But the plow-ahead gang pointed out the obvious: Obama would start the next Senate with fewer votes for ratification.

Biden was confident he could pull this off. The vice president had identified fourteen GOP senators who were in play and started meeting one-on-one with each.

"I always thought there would be enough Republicans who would think it's in the national interest to lower the number of nuclear weapons and who wouldn't want to damage the US relationship with Russia," a senior administration official recalled, "but that we wouldn't be able to show sixty-seven votes until close to the end."

THE GAME WAS STILL ABOUT KYL. DAYS AFTER THE ELECTIONS, Kerry, traveling in the Sudan at Obama's request, made a 3:00 A.M. call to Kyl to discuss the prospects for proceeding in the lame-duck session. The White House then committed to an extra $5 billion over the next decade for nuclear weapons modernization and maintenance. And in response to Kyl's request, Obama also produced the next year's budget for nuclear weapons activity months ahead of the normal schedule, so the senator could review it.

Biden and Secretary of State Hillary Clinton called Kyl to tell him about the new budget numbers—we think you'll be impressed, they said—and asked to send a team to Arizona to brief him. Two days later, the White House dispatched three officials to the Grand Canyon State—Jim Miller, the deputy undersecretary of defense for policy; General Kevin Chilton, the head of the Strategic Command; and Neile Miller, the deputy director of the National Nuclear Security Administration—to review the budget figures and discuss Kyl's never-ending concerns. Defense secretary Robert Gates also called Kyl. A meeting between Kyl and Clinton, Gates, and Biden was in the works.

The ratification effort was turning into a production that could be called *Waiting for Kyl*. Reid couldn't even consider scheduling a debate and vote until the White House reached an accommodation with him. Yet a deal did seem close at hand.

"Kyl never said to us, 'If you do all this, I'll let the Senate vote on START,'" a senior administration official recalled. "But it was implied."

Then Kyl pulled the rug out. On November 16, 2010, he issued a press release saying he did not believe there was time to consider the treaty in the lame-duck session, noting there were too many "complex and unresolved issues related to START and modernization." He said he looked forward to continuing to work with the administration on these matters—presumably in the next Congress.

Biden was presiding over the regular monthly meeting on Iraq in the Sit Room when news of Kyl's double cross reached the White House. The vice president gathered with McKeon; Tony Blinken, his national security adviser; Denis McDonough; and Ben Rhodes, the deputy national security adviser for strategic communications.

They quickly put out a statement in Biden's name declaring ratification could not be delayed. The point was to signal to senators that Kyl's sabotage was not the final act. Yet at this stage, the White House could count on only three Republican votes.

Reporters covering the treaty depicted Kyl's retreat as a deathblow. Some White House aides raised the possibility of bailing on ratification for the year. They didn't want to face a loss that would undercut the president, but that afternoon, Obama waved aside the questions and told his aides, "We've just got to go ahead."

OBAMA AND BIDEN HAD NOT ACTUALLY COUNTED ON KYL'S vote. "The vice president expressed more than once that we would never get Kyl," a senior administration official later said. The strategy had been to address Kyl's demands to keep him from raising a fuss that would cause other Republicans to shy away from ratification. If Senate Republicans thought Obama had treated Kyl shabbily, they would be more likely to side with him against ratification. Obama was willing to do much to make sure they had no such excuse.

"We wanted to show other Republicans we had gone the extra mile or five miles to satisfy Kyl," this aide noted.

Understandably, White House staffers and Kerry aides were

pissed off. Kyl had delayed the process repeatedly, with requests for information and assorted demands. Now he was saying they had to wait until next year?

But the White House quickly made a calculation: no one was to say anything nasty about Kyl. An assault on Kyl, Obama's aides thought, would prompt Senate Republicans—including those moderate GOPers on their wish list—to circle the wagons around him. When Robert Gibbs, during a briefing, was asked about Kyl's move, he calmly asserted that the White House would "continue to work through this process."

To outsiders, it may have seemed that Obama was refusing to fight tough. He had been sucker punched by Kyl, and the White House was barely responding. Here was more evidence for those who considered Obama a weak or uncertain combatant.

"In the old days," a senior administration official subsequently explained, "we'd have brought reporters into the Roosevelt Room and said—wink, wink—Here's what we're doing and why." Such an off-the-record chat would be held just so journalists wouldn't spread the impression that the president was shying from a fight. But these days, word of this sort of meeting would inevitably leak, and the White House's counter-Kyl strategy would become public. Restraint was a strategic resource.

This moment highlighted a source of frustration for Obama and his aides: the constant instant judgment of the pundits and the politerati. With the 24/7 cable-, blog-, and Twitter-driven media pouncing on every twist and turn of the main stories of the instant, Obama confronted dramatic pronouncements about his actions, decisions, and performance every day. He was up; he was down. This tactic guaranteed success; that move would ensure failure. Obama always counseled his aides not to be sucked into the cable-chatter black hole, even as he himself paid attention to how prominent commentators were scoring his daily efforts.

"We get graded every day," Axelrod noted (that is, complained). "But in this world, it's not how you do on the final exam; the media judgment unfortunately is an amalgam of daily quizzes. Yet Obama is focused on the exams. Worse, you get rewarded for the daily theatrics. And if you're not doing the daily thing viscerally, you're called detached. . . . You have to decide whether you do what you do to feel good at the moment or to do good overall. Plenty of times it's more viscerally satisfying to eviscerate the other side."

Obama, Biden, and their aides were opting against evisceration. "Our M.O. was always to tend to the long game," a senior administration official later remarked. "There often will be turbulence in the short run."

IT WAS TIME FOR OBAMA TO UP HIS OWN INVOLVEMENT IN THE New START battle. The White House scheduled an event at 1600 Pennsylvania Avenue—featuring former national security honchos and Republican luminaries backing the treaty—to rally support for the accord. Originally, Biden was to preside. Now Obama was handed the starring role to pump up the White House's PR push for ratification.

With Kyl now openly on the dark side, the White House and Kerry shifted their attention toward John McCain. He had griped about the treaty, but they thought he was reachable—and if he supported ratification, it would be easier for moderate Republicans to do so as well.

On November 18, after the White House event, Kerry hurried back to the Senate and arrived late at a meeting on the treaty with several Republican senators. The conversation was hostile. The GOPers besieged Kerry with questions: How can we ratify the treaty in such a short time? Doesn't it concede too much to the Russians? McCain looked like a human ball of anger. To one participant, ratification appeared to be a lost cause. But as they left the meeting, Kerry placed his arm around McCain, and the two whispered to each other. Moments later, Kerry told an aide: McCain has an amendment, and we need to work with him on it.

McCain's issue was the same one he had raised months earlier: the Russians declaring they would abandon the agreement if the United States went too far in deploying missile defense systems. Kerry knew this was a red herring because such declarations were common for arms control negotiations. But if he could satisfy McCain and gain his vote, that would be just as good as winning over Kyl. McCain might in the end be unattainable. But most of all, Kerry didn't want McCain, that old bull, leading a charge against the treaty. His goal was to string McCain along, keep him in play, and the White House was keen to do what it could on this front.

Obama and his aides felt they were on a roll. The White House event and a subsequent NATO summit in Lisbon (where the leaders of the NATO nations officially called for the ratification) appeared to generate momentum for the administration. But Biden, working through his GOP target list, was not obtaining firm commitments. His conversations convinced him that if the Republicans were provided a generous amount of time to debate ratification on the Senate floor, there would be enough votes to pass the treaty. Yet some White House aides wondered if Biden was overstating his case—and overestimating his own powers of persuasion.

ON DECEMBER 15—THE DAY THE SENATE PASSED THE TAX-CUT compromise—the upper chamber voted to begin debate on ratification. The day before, Kyl and a group of influential Republican senators had held a press conference to denounce the Democrats for rushing ahead with ratification. (This rushing charge was thin; there had been a half year of hearings and ceaseless back-and-forth with Kyl and others.) Kyl went so far as to accuse Obama and Reid of "disrespecting one of the two holiest days for Christians" by debating the treaty so close to Christmas. Republican Senator Jim DeMint declaimed the situation as "sacrilegious."

Kyl, though, still refused to say, after all this time, if he would vote against ratification. "If I announce for or against the treaty at this point, nobody would listen to me," he said, acknowledging his own gamesmanship.

Reporting on this press conference, the *Wall Street Journal* called it a "decisive blow" to the treaty.

Kerry's task was to manage a floor debate that afforded the Republicans room to vent while smothering (oh so respectfully) the dozens of amendments they were proposing. If an amendment to alter the treaty were to pass, the accord would have to be renegotiated with the Russians—which would threaten the entire agreement.

At an "emergency" meeting of the Democrats of the Foreign Relations Committee, Kerry counseled self-control: "Don't go to the floor and say anything angry about the Republicans. It doesn't get us one more vote."

The White House, with Brian McKeon in the lead, set up a war room in the Foreign Relations Committee's ceremonial room on the first floor of the Capitol and staffed it with the lead negotiators and experts from the State Department, the Pentagon, and the Department of Energy. They were quick to whip up talking points, respond to amendments and assertions from Republicans, and answer technical questions posed by senators.

Over several days, there would be seventy hours of debate— far more than other arms control treaties had drawn. For most of this time, Kerry was at the center of the floor action, parrying with Republicans who raised objections or offered killer amendments. He maintained his cool through it all—and kept the partisan temperature low—to smooth the way for moderate Republicans to support the treaty.

Biden and Kerry were doing whatever they could to pick up Republican senators one by one. Kerry arranged for Democratic leaders in the Senate to cosign a letter to Obama requested by two Republicans—Thad Cochran and Lamar Alexander—asking the White House to treat the nuclear weapons budget (housed within the Department of Energy) as a military line item. This would protect the program from nondefense spending cuts. Obama agreed to do so—and pocketed these Republican votes.

Biden kept talking to McCain, trying to work out something—or at least to prevent McCain from declaring his opposition to ratification. (Around this time, McCain and Lindsey Graham were letting the White House know that if it abandoned the effort to repeal Don't Ask/Don't Tell, they could be convinced to vote for ratification.)

On December 16, Kerry handed McCain a letter from Obama that was meant to address McCain's key objection; it reiterated the White House position that the New START treaty would place no limits on US missile defense programs. The following day, McCain, with Graham by his side, came to the White House to talk with Obama and Biden. The meeting was not placed on the president's public schedule.

ON THE MORNING OF DECEMBER 18, A SATURDAY, KERRY WAS AT the White House to have breakfast with Donilon and plot final strategy. Donilon was running late, and while Kerry waited, he called Reid. There was a problem: McCain.

Reid was considering later in the day filing the cloture petition that would set up a vote to end debate and move to a final vote on ratification. The White House had not yet signed off on that, but this was truly not the right time for Reid to take this step.

Later that morning, the Senate was scheduled to vote on the repeal of Don't Ask/Don't Tell. The White House had the votes to win that hot-button face-off, and there was tremendous anger among Republicans on the losing side. Before Kerry had reached the White House, he had caught wind of the fact that McCain was livid about the cloture petition for the New START bill—and McCain was already irate about the pending repeal of Don't Ask/Don't Tell.

If Reid moved this day to end debate on the treaty, McCain might be pushed over the edge. To some that might sound silly: the fate of an agreement governing nuclear weapons dependent on the mood of a resentful legislator. But that was how the Senate worked. Legislative decisions were not always based on policy or politics; sometimes it was personal.

"We feared an avalanche if McCain said no that day," a Senate aide recalled. "We thought he was the linchpin."

Though McCain had continued to be grumpy about the treaty, Kerry and his aides still believed they had a chance of roping him in for the final vote. But now their main goal was to keep him out of the nay column—if only for a day or two. Several of the moderate Republicans on their wish list would be lost if McCain were to go ballistic.

Kerry reached Reid on the phone: "Harry, don't do this now. We'll lose."

The Senate majority leader saw the logic, and a potential treaty-busting explosion was averted. Later that day, during the continuing floor debate on the treaty, Kerry read the letter on missile defense Obama had written for McCain. The Arizona senator also was given the chance to offer an amendment that would excise language from the treaty's nonbinding preamble noting a relationship between offensive and defensive weapons.

(McCain claimed that this portion of the treaty would deter US plans for missile defense—a contention challenged by arms control experts.)

McCain's amendment was shot down on a 59–37 vote. But Kerry and his aides talked with McCain about alternative amendments in which the former Republican presidential candidate could express his sentiments regarding missile defense. The goal: keep McCain on the hook.

THE WHITE HOUSE COULDN'T WAIT ANY LONGER TO MOVE toward a vote. The Senate would be adjourning in days. The next day was the deadline for filing a cloture petition that would allow enough time for a final vote.

That Sunday morning, McConnell declared his opposition to ratification, and Kyl finally did the same. Biden was now confident he had at least thirteen GOP votes—more than enough to hit the sixty-seven mark. But he wasn't sure which Republicans. A senior Democratic Senate aide estimated that Kerry could depend on about sixty-four or sixty-five senators to back ratification, with several still up for grabs. Close but no guarantee. That night, Reid, per Obama's decision, filed for cloture. The vote would happen in two days.

"The president made a gutsy decision that he was willing to lose it," Kerry later said.

Then things fell into place for the White House. Kerry had been leaning on Senator Scott Brown, the Republican from his home state of Massachusetts, and Brown announced he'd support the treaty. The White House arranged for Mullen to appeal to Republican senators, and other Republicans either endorsed ratification or signaled they were leaning in favor: Judd Gregg of New Hampshire, George Voinovich of Ohio, Bob Bennett of Utah, and Olympia Snowe of Maine. Counting the votes of Lugar, Corker, and Cochran, the White House was almost there.

"My hat's off to the Democratic leadership," Senator Lindsey Graham said. "They're running rings around us. . . . They're like Sherman going through Georgia."

The morning of December 21, Biden was busy calling Republican senators to reaffirm they'd be voting for cloture. At a press conference, Kyl moaned that "this process has not enabled us to consider this treaty in the serious way it should have been considered."

It was another hollow charge: the Senate was devoting a full week—including a weekend—to debating the treaty. (The original START treaty had received only five days; START II, only two days.) Graham decried fellow Republicans for acting as accomplices as the White House "jammed" the accord through the Senate.

Later that day, the Senate voted 67–28 to end debate and proceed to the ratification vote. Eleven Republicans had joined the triumphant side; McCain had not. Obama and Biden had overcome the obstreperous opposition led by Kyl and McConnell (while neutralizing McCain) and forged a bipartisan coalition.

The final vote the next day was almost anticlimactic. As Biden presided over the Senate and Obama watched on a small television screen in the Outer Oval Office, the treaty was ratified on a 71–26 vote, with the White House picking up two more Republicans. Biden, Kerry, and Lugar had persuaded almost one-third of the Senate Republican caucus to break with its leaders. The sophisticated and deft strategy devised by the White House and Kerry had resulted in a complete success.

The treaty was a modest one, yet overcoming the Republican defiance was an important victory for the president. "No Russian-American arms treaty submitted for a Senate vote ever squeaked through by a smaller margin," the *New York Times* reported. "But for a president seeking his way after a crushing midterm election, it was a welcome validation that he could still win a battle."

The Hot Air conservative website offered a dispirited but snarky reaction: "The tax cuts deal, then DADT repeal, now this. Man, I'll bet The One wishes every session could be a lame-duck session."

SHORTLY AFTER THE SENATE VOTE, OBAMA WALKED FROM THE West Wing to the Eisenhower Executive Office Building to hold a press conference. It would be his last before the new political reality hit Washington. As he strode across the driveway between the two buildings, Gibbs, Axelrod, and Pfeiffer were with him, each of them smiling.

At the podium, Obama evinced triumph. After the midterm elections, he said, "A lot of folks in this town predicted Washington would be headed for more partisanship and more gridlock. And instead, this has been a season of progress." And there was more than the tax-cut deal, New START, and the repeal of Don't Ask/Don't Tell. Obama cited a food safety bill ("the biggest upgrade of America's food safety laws since the Great Depression") enacted in the lame-duck session and pointed out that Congress that day was on the verge of approving legislation previously blocked by Republicans that would pay for medical care needed by 9/11 rescue workers. Congress had also okayed a child nutrition bill that expanded the school lunch program—a bill much favored by Michelle Obama.

"This has been the most productive postelection period we've had in decades," Obama said.

He was correct. After suffering a historic electoral loss, he had scored a series of notable policy victories.

—᷄᷄᷄—

THERE HAD BEEN PLENTY OF SKEPTICISM INSIDE AND OUTSIDE the White House after the elections. "Nobody in the West Wing thought he would get as much as he got," Bill Burton recalled. But Obama had not yielded to the nay-saying within his own circle.

"Being a great leader means absorbing all the anxiety and doubt and knowing in your own heart your commitment," a senior administration official said of Obama. "He absorbs a lot of pain."

In the ten weeks since the "shellacking," Obama had demonstrated a political dexterity, outflanking fierce Republican opposition on several fronts. The rap on him had been that he was a lousy negotiator—and he might indeed have conceded too much during the fights over the stimulus and health care—but in the recent weeks he had employed a firm, though quiet, negotiating style that allowed him to outmaneuver Republicans. The downside: much of this transpired far from public view. Obama had achieved these wins not by mobilizing millions and shaping the overarching narrative but by engaging in crafty negotiations and tactics.

Nevertheless, as White House aides saw it, Obama had boosted his appeal for various constituencies. The tax-cut deal had impressed independent voters. The repeal of Don't Ask/ Don't Tell was a reminder for die-hard progressives. The New START treaty ratification played well with the keepers of elite opinion.

WITH THIS LAME-DUCK SESSION OVER, THE POLITERATI WERE looking toward the next act: Obama versus the Republican House. Could he negotiate with the Tea Party–driven GOPers?

Was he prepared to duke it out with them? How could he use this coming year to set up what was already looking to be a damn difficult reelection?

When reporters at the press conference asked him about the prospects of working productively with Republicans in the months to come, Obama optimistically declared it would be possible to find "common ground." The Republicans, he predicted, will "recognize that with greater power is going to come greater responsibility."

Maybe. Obama appeared to be in a charitable—perhaps too charitable—mood. The day before, Congress had approved a temporary funding bill for the federal government, averting a shutdown. The law would keep the government afloat only until March 4, setting up a full-fledged budget battle early in the new year.

That evening, Obama shared a champagne toast with White House staffers who had worked on New START. Then he departed for a family vacation in his native Hawaii. He knew he would soon be returning to a much different and more difficult Washington.

Yet as he boarded Air Force One, the president was smiling and singing "Mele Kalikimaka," the popular Hawaiian Christmas song.

Chapter Five

THE STRATEGY

A DEWY-EYED JOHN BOEHNER, THE SON OF A CINCINNATI saloon owner, stood at the podium of the House of Representatives on January 5, 2011, and took the gavel, becoming the sixty-first speaker of that body. Referring to the midterm elections, he proclaimed, "The people voted to end business as usual." He nobly added, "We will welcome the battle of ideas, encourage it, and engage in it."

President Barack Obama, back in the White House following his family vacation in Hawaii, was awaiting that battle of ideas. He had a plan for the next phase of his presidency—and it entailed precisely that: a clash of visions.

Obama and everyone else knew what the Republicans would be selling: small government, cutting taxes, and fewer regulations on business. They would be blaming the government for the country's woes, demeaning communal action, and insisting that the American people and businesses could right the nation on their own, if only the bureaucrats and busybodies of Washington would get the hell out of the way.

Obama had a different take. He saw the great American

middle class being hammered by global and domestic forces that individual families and businesses could not counter on their own. To face competition from China and other emerging economic superpowers, the nation required forward-looking and proactive government policies that would guide Americans through the rough transition under way.

The current economic malaise had to be addressed. The unemployed needed assistance. Jobs had to be created. Consumer and business confidence had to be boosted. The housing market had to be bolstered. The financial sector had to be monitored closely. Teachers, firefighters, police officers, and others had to be kept on the job. But Obama also believed he had to prep the country for the years ahead, seeding and encouraging innovation and new industries, advancing research and development, reviving the education system, and shoring up the social safety net.

This guaranteed another round in the historic tussle over the role of government in American life—the sort of grand and foundational conflict perfect for a presidential election. But before Obama could have this debate with the Republicans, he had to address a debate within the White House, which was under new management.

IN THE WEEKS AFTER THE NOVEMBER ELECTIONS, OBAMA mulled over how he could have a more efficient and effective White House. He knew there was certainly room for more discipline. "He wanted more rigidity and more sticking to the plan," a senior aide later recalled.

That meant more control from the top. So as his new chief of staff, he brought in William Daley, the Midwest chairman of JPMorgan Chase. Pete Rouse, who had well managed the White House as the temporary chief of staff during the intense lame-duck session, became counselor to the president.

Daley, who had been a secretary of commerce during the Clinton years and a chairman of Al Gore's presidential campaign, straddled the financial and political worlds. He had been one of the first supporters of Obama's presidential bid (breaking with the Clintons to back his fellow Chicagoan). He had co-chaired the transition's economic policy department.

Obama's decision to enlist a senior executive of a major financial firm did not kick up much of a fuss. Though JPMorgan Chase had retained an army of lobbyists to beat back Obama's far-from-radical Wall Street reform bill—and to oppose the creation of the consumer financial protection unit—Obama did not hold that against Daley, whose brother was the outgoing mayor of Chicago. (Daley, who had helped oversee the bank's lobbying operation, would recuse himself from the deliberations concerning whom to select as head of the Consumer Financial Protection Bureau.)

By picking a Big Finance guy—Daley even bore a passing resemblance to Daddy Warbucks—Obama was indeed sending a symbolic tweet to Wall Street and the business community. It was, as an Obama adviser put it, a signal that "there are more open doors here, come and have a conversation." Obama had not stopped fretting about the antipathy toward him within the corporate class. (In mid-January, he would unveil the President's Council on Jobs and Competitiveness—and name General Electric CEO Jeff Immelt as chairman.)

Obama wanted strong executive-style management at 1600 Pennsylvania Avenue. "Whichever way this president wants to go, Bill is gonna make sure that it's done in the most effective way possible," Chris Lehane, a Democratic operative who worked with Daley on the Gore campaign, told *New York* magazine. But, as one Obama adviser noted, "Bill certainly doesn't share the president's political values. Then again, Rahm didn't either. It might be good for him to have a chief of staff who did."

Daley desired a coherent game plan—and a more organized game. He preferred an orderly and hierarchical decision-making process. He immediately canceled the daily 8:30 A.M. gathering of heads of White House departments. He dramatically constricted who could attend meetings. Staffers grumbled about this, but Daley wanted the circle to be tighter. That would at least reduce leaks. Emanuel used to walk the halls of the West Wing, yanking people out of meetings to deal with the urgent matter of that instant. Now Daley was in his office, often in one scheduled meeting after another, with the door closed.

As a onetime Washington insider, Daley knew how to deal with the main players of the town. (The denizens of Washington's salons had long griped that Obama hadn't sufficiently courted the town's bipartisan establishment.) The new chief of staff had a positive relationship with John Boehner. And for his first meeting with Richard Trumka, Daley walked across Lafayette Park to the AFL-CIO headquarters and spent ninety minutes with the union president. "Emanuel would have never done that," a labor official remarked.

DALEY'S ARRIVAL IN EARLY 2011 COINCIDED WITH THE RETURN of David Plouffe. After managing the grueling 2008 campaign for Obama, Plouffe had stepped back to write a book, renew his relationship with his family, and recharge. (He also consulted for General Electric and Boeing.)

Plouffe and Axelrod had earlier arranged a deal: Axe would serve in the White House for the first two years and then a revitalized Plouffe would sub in for him, and Axelrod would return to his family in Chicago and work on Obama's reelection campaign (which Messina was managing) as the guardian of the Obama message.

Plouffe and Axelrod had been partners in a political consulting business prior to the Obama campaign. Axelrod was the concept guy in charge of the grand themes; Plouffe was the hands-on manager overseeing the inch-by-inch ground operation. Axelrod did the poetry; Plouffe supplied the bone-crushing discipline. He was methodical and mechanical.

During the 2008 primary campaign, Plouffe obsessively fixated on one fundamental: the delegate count. And throughout the entire year, he crunched data. He pored over spreadsheets. He was the Joe Friday of politics: *just the metrics.* He never got too high or too low. "Bizarrely stoic," a campaign aide once called him. Another Plouffe colleague observed, "He's not dissimilar from the president this way." A senior official said of him, "I can't tell by looking at him what's getting him excited."

According to Plouffe's campaign memoir, *The Audacity to Win,* his experience managing that historic endeavor had reinforced for him the ultimate importance of envisioning a goal, developing a plan to reach it, and *sticking to the plan.* Plouffe had once been and was now again the keeper of the Obama strategy.

The timing of the Plouffe-Axelrod trade was good for Axelrod. He was exhausted. And as he left, he realized the White House did need change: "We were so enmeshed with trying to solve some significant challenges that we didn't get to tell a story in the first couple of years."

Robert Gibbs would also be leaving the White House—off to rejuvenate and then perhaps join the reelection campaign in some capacity. Having devotedly served Obama as an adviser on politics, policy, and messaging since Obama's successful 2004 Senate campaign, he possessed the institutional memory. He could recall what Obama had said about a certain trade policy six years ago during a Senate floor debate. He was as responsible as anyone for guiding and shaping Obama's meteoric ascent. But his stint as Obama's ambassador to the media had not always been smooth.

"Obama wondered if Gibbs had poisoned the relationship with the press," an Obama adviser said.

Gibbs's successor was Jay Carney, a former Washington bureau chief for *Time* magazine who had traded journalism for politics by becoming Biden's press secretary two years earlier. Carney, who would assume the press room podium in early February, could be expected to play—and parry—more nicely with the journalists. With Gibbs departing, Dan Pfeiffer would take on a more centralized role as communications director.

Now that Daley and Plouffe were in charge, a White House aide said, there would be "less flying-by-your-pants."

OBAMA HAD ANOTHER IMPORTANT PERSONNEL DECISION: WHO would replace Larry Summers as director of the National Economic Council? Some of the president's advisers, always looking to earn Obama corporate-class cred, suggested he bring in an outsider with business ties. Obama had done that with Daley. For this post, he chose Gene Sperling, an adviser to Geithner who had held this position during Bill Clinton's second term.

Sperling's selection was slammed by several Washington progressives who, citing his ties with the much-maligned Goldman Sachs investment bank, complained that Obama had recruited another Wall Streeter. But Sperling had been retained by the Goldman Sachs Foundation as an adviser to its $100 million program to elevate the skill levels of poor women in developing nations. Prior to that, at the Council on Foreign Relations, he created and led the Center for Universal Education, which promoted the need for educating young girls in developing nations. He was the rare senior economic official who had not fully cashed in after leaving government service. (He did well enough, earning nearly $900,000 from Goldman Sachs for this charitable project.)

During the Clinton years, Sperling had been a strong advocate of the earned income tax credit, a big gain for low-income working Americans, and a champion of increased spending for Head Start and other programs for children. He promoted debt relief and debt cancellation for poorer nations. He had been part of an economic team—headed by Summers—that pushed for financial deregulation but was not deeply involved in the effort.

Sperling was something of a neoprogressive. He had once chastised a group of labor-oriented, liberal economic policy advocates he dubbed the "Three Bobs"—Robert Reich (for whom Sperling once worked), Robert Kuttner, and Robert Borosage—for not caring sufficiently about budget deficits.

At the same time, though, Sperling advocated policies with progressive goals: retraining workers years in advance of expected dislocations (due to globalization and trade compacts) and establishing a universal government-subsidized 401(k) plan that would augment Social Security, not replace it. He called for estate taxes on the rich to pay for this retirement plan. In other words, he favored redistributing the wealth.

He considered himself a progressive but a realist when it came to working with Republicans. He would tell people—and Sperling was always eager to explain this point—that his desire to win results through negotiations (and compromise) was not a product of political centrism; it was pragmatism. In this way, he matched the president's own sentiments.

Sperling had lived through President Clinton's ferocious budget battles with Newt Gingrich and the Republicans, resulting in a government shutdown widely perceived as detrimental to Gingrich and his comrades. Budget director Jack Lew was a veteran of those engagements, too. The two shared similar scars and worked well together. (Summers and Peter Orszag had often clashed.) And Sperling got on well with Geithner, his boss at Treasury. Obama's economic team 2.0 would be less dysfunctional than the initial squad but forced to create policy within a tougher political environment.

One of Sperling's selling points for Obama was his experience in that zone where policy and politics overlapped. With Sperling, Obama was getting the economic policy equivalent of a wartime consigliere.

THE PRESIDENT AND HIS REVISED BAND OF ADVISERS HAD TWO main big-picture items of business: the State of the Union speech on January 25 and the submission of the next year's budget. Though much of the budget had been hammered out, Obama needed to decide on the overall message of the speech and the budget plan.

The Tea Party–fueled House Republicans had roared into Congress and appeared to own all the political momentum, but the State of the Union and the budget plan would be Obama's shot at shaping the national agenda. At the least, he could stake out his own turf.

White House aides discussed whether Obama should have an era-of-big-government-is-over moment. That's what Bill Clinton had proclaimed in his 1995 State of the Union speech, reacting to the Gingrich-led GOP House takeover of 1994. It had been Clinton's way of absorbing the dramatic change in the political landscape. The Obama White House was loaded with Clinton administration veterans, and that example was on people's minds.

Obama—no surprise—wasn't interested in a Clinton-like pirouette. He told aides he wouldn't engage in any grand gestures, and he certainly would not backtrack on health care reform and his other key initiatives. He would adapt to the new circumstance, yet not retreat. He would focus more on persuasion. But what ought to be the subject of presidential persuasion?

In December and January, the White House held a series of

strategy meetings. The core question was whether the president would embrace deficit reduction as a—or *the*—top priority. Over several Saturday mornings, Daley; Plouffe; Pfeiffer; Jarrett; Sperling; Schiliro; Nancy-Ann DeParle, deputy chief of staff; Melody Barnes, director of the Domestic Policy Council; Bruce Reed, who replaced Klain as Biden's chief of staff; and other aides gathered around the long wooden table in the windowless Roosevelt Room. Geithner, Lew, and Axelrod attended some of the sessions.

Obama's brain trust was peering ahead and trying to figure out how the next six to eight months would play out. They anticipated that the debt ceiling would be a struggle. They also concluded that Obama, as one participant recounted, would have "to show some leg on deficit reduction."

In early December 2010, the Bowles-Simpson deficit-reduction commission Obama created the previous February had released its report. It called for blotting up $4 trillion in red ink over the coming decade, with provisions that would upset the Left and the Right: raising the age for Social Security (while protecting low-income beneficiaries), cutting Medicare by about $400 billion, trimming $1.6 trillion in government programs (both military and nondefense), and raising tax revenues, while reforming the tax code and lowering rates.

The commission also backed a shot of stimulus for 2011, and its report insisted that budget cutting not begin until 2012 and then proceed gradually, so this would not interfere with the slow recovery. Overall, the commission called for a three-to-one ratio in spending cuts to revenue boosters.

The centrist pundits and pols of Washington, a group long yearning for deficit reduction, hailed the plan's arrival. But for the report to be officially handed off to the president and Congress, fourteen of the eighteen commission members had to vote to accept it. Yet seven said no. All three of the House Republicans on the panel—Paul Ryan, Jeb Hensarling, and Dave Camp— voted thumbs down, complaining that the report didn't undo Obama's health care overhaul and insisting that they couldn't support any tax hikes. Nevertheless, the Obama-created commission placed pressure on the president.

In the weeks after the report's release, Obama's aides debated how best to respond—and these discussions continued into the strategy sessions of the new year. Citing the perils of deficits and debt, the new GOPers were poised to wage a slash-and-burn campaign over government spending. And everyone in the White House agreed that Obama should find a seat on the antideficit train—if for no other reason than to win over voters to his other economic policies.

"If you can instill some confidence in the public mind that you can deal with these deficit issues," a senior administration official later said, "you have more room in the short term to promote your favorite policy matters."

Plouffe was concerned that voter unease about the deficit could become unease about the president. The budget issue was easy to understand: you shouldn't spend more money than you have. Yes, there was the argument that the government should borrow money responsibly when necessary (especially when interest rates were low!) for the appropriate activities, just like a family borrowing sensibly to purchase a home, to pay for college, or to handle an emergency. But voters needed to know—or *feel*—that the president could manage the nation's finances. The budget was a test of government competence—that is, Obama's competence.

LINES WERE DRAWN WITHIN THE ADMINISTRATION OVER whether Obama should enthusiastically embrace deficit reduction as his top cause and offer a specific and extensive budget-shrinking plan to counter the on-the-rampage Republicans, or devote more energy to promoting a "growth agenda," a series of future-oriented investments in education, innovation, and infrastructure.

Geithner and other advisers seized upon the Bowles-Simpson report as a signal that "deficit reduction ought to be our number one priority," a former senior administration official recalled. This would certainly win Obama kudos within the Beltway, for there was never-ending pressure from the centrist Democrats and the Washington establishment for deficit reduction.

But the political team in the White House was not convinced. Neither was Austan Goolsbee, who had succeeded Romer as chair of the Council of Economic Advisers. Goolsbee and others favored pushing this growth agenda that included programs and policies expected to yield economic expansion. They believed Obama needed to espouse a forward-looking and upbeat view that emphasized the nation's potential to meet the economic challenges of the coming years. In bumper-sticker talk: Educate, Innovate, Build. Goolsbee, for one, feared that a deficit-heavy State of the Union speech would undercut this more positive message.

Sperling thought it would be idiotic for the president to adopt the Bowles-Simpson proposals as his opening position, for Republicans would only insist on more cuts and fewer revenues. It might be fine to end up near the commission's recommendations, but presenting them as a first-round bid would only drive the process farther to the right.

The State of the Union—the embodiment of the president's message—would definitely cover both deficit reduction and the growth agenda. But what would be the balance? As one of the senior aides later noted, "The question was, what's the full meal and what's the dessert?"

THAT DECISION WOULD BE RENDERED WITHIN A LARGER STRATE-
gic context because Plouffe wanted to trap the Republicans.

His goal was to orchestrate circumstances so that the GOP
would release a budget plan targeting Medicare and other entitlements (such as Medicaid and Social Security) before Obama
laid out his own deficit-reduction proposals. Representative Paul
Ryan, the Republican now chairing the House Budget Committee, had previously unveiled a blueprint calling for replacing Medicare's guaranteed coverage with something akin to a voucher—a
change that would force seniors to pay more for health care. If
the House Republicans voted for any budget legislation resem-
bling Ryan's scheme, Obama and the Democrats could have a
field day. This was Negotiations 101: make the other guy go first.

Some administration officials argued that Obama would be
derided if he didn't produce a serious deficit-busting plan of his
own. And to be deemed serious such a plan would have to address Medicare's rising costs. Deputy chief of staff Nancy-Ann
DeParle countered that it would be foolish to propose any specific Medicare reductions before the Republicans had a chance
to hoist themselves with Ryan's petard. Which was what Plouffe
was anticipating.

Plouffe, the master tactician, wanted to hang back—and wait
to see the whites of their eyes.

THE PRESIDENT HAD CONCLUDED THAT SOME VOTERS WERE
riled up by all the government spending of the past two years
and needed to be reassured about the deficit. He also believed
that addressing the government's debt load was necessary for the
sake of the economy. It was not only politics. The economic indicators and forecasts of the moment showed glimmers of improve-
ment. Yet the markets seemed concerned—however that could
be determined—by the large deficits and growing national debt.

"A big cloud over the economy was the issue of debt," Axelrod later said. "You heard that from the analytical world. Dealing with it was part of the economic strategy, not just a matter of choosing the best message."

But Obama didn't want to get mired in the hardly inspiring arena of deficit reduction. He aspired to present a more exciting and engaging economic vision that extended beyond the made-for-accountants nitty-gritty of budget crunching. His idea was to shift the political focus to the future by hitting the interconnected themes of growth, competitiveness, education, and innovation. Promoting these notions, he could even advance policies that Republicans would accept—or, at least, that they had in the past.

"He wanted to have a real debate on the future of the country," Gibbs recalled. "He could not envision a scenario where innovation and investment would be superseded by debt and deficit. It was the president's view that people understood we had been on a slow journey of becoming less competitive and that they wanted a long-term economic plan that helped in the short term. Not just CPR; there had to be long-term wellness."

Obama decided he would use the State of the Union to tell the story of how the nation could grow out of the hole it was in over the long run. Tethered to that tale would be policies that in the past had not been considered highly partisan: funding for research, education, and infrastructure.

Meanwhile, Obama, as Plouffe advised, would not make the first move in the deficit face-off. He would wait for the GOP to produce its own (no doubt, extreme) plan, let it be chewed up, and then strike. This would irritate the budget hawks of Washington; they could be counted on to accuse the president of a failure in leadership. But the president would just have to take that arrow.

Left out of the equation were any immediate jobs programs.

Unemployment had dropped ever so slightly the past few months. Forecasters were predicting modest growth ahead. Obama and his advisers did not see this as a moment to push another shot of government spending for a here-and-now jobs initiative—especially so soon after turning the tax-cut deal into a second stimulus. They would lead with what one White House official subsequently called "less of a jobs message."

After all the discussion and wrangling, Obama and his team resolved that with the State of the Union and budget plan, the president would seek to transcend the immediate confines of the ongoing political and policy debates. Axelrod, Plouffe, and Pfeiffer had concluded that for the past two years the American public had seen Obama contending with and bearing the burden of crisis—day in, day out. Now was a moment when he could go bigger and convey his ideas about restoring the American Dream—and position himself for a confrontation with the Republicans over basic values and vision. That was the plan.

THE REPUBLICANS WERE ALSO PLOTTING STRATEGY. IN MID-January, the House Republicans gathered at the Marriott in Baltimore's Inner Harbor for a three-day retreat behind closed doors. In private sessions, the GOPers, including the eighty-seven newcomers who had reached Congress surfing the Tea Party wave, listened to Governors Rick Perry of Texas and Haley Barbour of Mississippi and were schooled in economic policy by CNBC host Larry Kudlow and former Reagan adviser Arthur Laffer. Perhaps the most important moment came when the Republican House majority leader, Eric Cantor, addressed the flock.

Cantor was the renegade prince of the Republican conference. Boehner, his supposed boss, was conservative, but more of a let's-make-a-deal-and-then-hit-the-golf-course legislator of an earlier era. He was not an ideological warrior driven to the barricades by the rise of Obama. Boehner, though, had become speaker only because of the influx of extremists into his House and couldn't forget that. Yet Cantor was closer in style and views to the newcomers. More important, he was ambitious, and he saw there were troops to be led.

Cantor reminded the Republicans that in a few months, the federal government would reach its $14.3 trillion debt limit. This would be, he said, a "hidden" opportunity, and they would have the chance to dictate the terms.

This was a much different message from what Boehner had conveyed two months earlier to the eager-to-slash freshmen.

In a mid-November meeting with reporters, Boehner had said, "I've made it pretty clear to them that as we get into next year, it's pretty clear that Congress is going to have to deal [with raising the debt ceiling]. We're going to have to deal with it as adults. Whether we like it or not, the federal government has obligations, and we have obligations on our part."

Boehner had been living in the world as he had always known it, sharing the traditional—and responsible—view that you don't monkey around with the debt ceiling and default. And the White House took his reasonableness seriously—maybe too seriously.

Cantor was steering his new colleagues into a new world. The old ways didn't matter; they needed shattering. And the Tea Party members' primary cause was to decimate the bloated federal government. By any means necessary. In this hotel conference room, Cantor was telling them how to do it.

"I'm asking you to look at a potential increase in the debt limit as a leverage moment when the White House and President Obama will have to deal with us," he explained. "Either we stick together and demonstrate that we're a team that will fight for and stand by our principles, or we will lose that leverage."

It was a polite way of saying, "Our hostage taking will be

more successful if we remain united." Cantor was counting the days until he and his band would have Obama in a corner.

AFTER MIDNIGHT, IN THE EARLIEST MINUTES OF JANUARY 12, 2011, Obama was rewriting the draft of a speech that had been forwarded to him several hours earlier. He had expected to finish the revision by 11:00 P.M. But the midnight hour came and went, and then 1:00 A.M., and then . . . at 1:20, he sent the reworked draft to Axelrod to review.

Obama's main message man saw that the boss had revised the final passages of the speech to boost its call for civility in the national political discourse. Obama had been up late striving to turn a tragedy into a teachable moment.

On January 8, a disturbed young man named Jared Lee Loughner arrived at a Tucson Safeway supermarket, armed with a Glock 19 semiautomatic pistol and an extended magazine of ammo that afforded him increased firepower. Representative Gabrielle Giffords, a Democrat, was holding a meeting with constituents at the site. Loughner shot her in the head and then proceeded to fire at others before brave bystanders wrestled him to the ground. Giffords survived but sustained serious brain injuries. Six people were killed, including Christina Taylor Green, a nine-year-old girl.

Though Loughner's motives were unclear, the massacre sparked fierce commentary about the harsh tone of the political debate. After all, months earlier, Giffords had pointed out that Sarah Palin's political action committee had targeted Democratic legislators it hoped to defeat—including Giffords—on a map by placing crosshairs over their congressional districts.

"When people do that," Giffords said (after her office was vandalized), "they've got to realize that there are consequences to that action."

In the uproar following the shooting, Palin responded to critics in a moment of abject self-absorption by claiming she herself was a victim of a "blood libel."

Obama was asked to speak at the memorial service to be held four days after the tragedy. He pondered what message to send to a nation frozen by the Tucson horror. Should he address the topic of gun violence? Use the moment to chastise the practitioners of overheated rhetoric?

THAT DAY, HE STOOD AT THE PODIUM IN AN AUDITORIUM AT THE University of Arizona and spoke as the preacher in chief.

"There is nothing I can say that will fill the sudden hole torn in your hearts," he told the crowd, which moments earlier had cheered him loudly. He paid tribute to the fallen and to the heroes and survivors of that awful morning. He noted that this tragedy, coming during a time of "polarized debate," should cause all Americans to "pause for a moment and make sure that we're talking with each other in a way that heals, not in a way that wounds."

He pleaded for "humility" in public discourse: "Rather than pointing fingers or assigning blame, let's use this occasion to expand our moral imaginations, to listen to each other more carefully, to sharpen our instincts for empathy and remind ourselves of all the ways that our hopes and dreams are bound together."

A president who had been accused by political foes of being a secret Kenyan-born Muslim and/or socialist who yearned to create death panels and destroy the economic well-being of the nation was issuing a reasonable cry for respect, civility, and unity, even as contentious and irreconcilable differences divided the political class.

Toward the end of the address, Obama noted that victim Christina Taylor Green had been born on September 11, 2001, and featured in a book about fifty children who had entered the world that day. In that volume were simple wishes for a child, including "I hope you jump in rain puddles."

Obama continued: "If there are rain puddles in heaven, Christina is jumping in them today. . . . And we commit ourselves as Americans to forging a country that is forever worthy of her gentle, happy spirit."

It was a powerful performance. The president made it feel—at least for a few minutes—as if the United States was a small town, with every resident touched by this ghastly event.

The president walked back to his seat in the front row. He kissed and hugged his wife. Michelle Obama was holding back tears.

Obama, bracing for a tough stretch of political combat this year, had done what he could to reset the national tone. He had not used the tragedy to advance any particular policy. Afterward, conservative columnist Charles Krauthammer praised the speech and said he "wouldn't underestimate how this is going to affect the perception of the president." But while the moment seemed rich with possibility, Obama had to realize that a single tragedy would not yield a more peaceful and productive environment for the unavoidable conflicts ahead. That would be too much to wish for.

A WEEK LATER, THE NEWLY IN-CONTROL REPUBLICANS TOOK their first legislative action. It was no olive branch. Fulfilling a prominent campaign promise, they voted unanimously to repeal Obama's health care overhaul, without proposing any replacement measures. In the aftermath of the attempted assassination of Gabby Giffords, the floor debate had been slightly subdued. Still, Republicans had decried Obama's health care reform as "socialistic" and a government takeover (which it was not). With the bill DOA in the Democratic-controlled Senate, White House aides did not worry much.

"This isn't a serious legislative effort," Robert Gibbs, still in the job, told reporters.

And it wasn't. The Republican move was more a perfunctory exercise to cross an item off their to-do list. Perhaps that was because recent polling had indicated that both opposition to the law and support for repeal had fallen. For now, the most important piece of business for them was their campaign promise to slash $100 billion in spending. That was the battle they were preparing for.

ONCE THE PARAMETERS OF OBAMA'S STATE OF THE UNION HAD been decided, the speech needed a theme. Jon Favreau, the president's chief speechwriter, cooked up "Winning the Future."

This rhetorical flourish was, an outside Obama adviser noted, "a little forced." But it tied a bow around Obama's priorities and his pro-growth message. It conveyed optimism. The Republicans would be pushing an eat-your-spinach message of cuts and more cuts. Obama had to trump their worst-of-times carping with a more encouraging refrain that encompassed more than deficit bean counting. And Plouffe saw how this could be an organizing theme for the arduous year facing the White House.

Axelrod and others at the White House had been heartened to see the president's approval rating tick up after the lame-duck session—especially among independents. "The image of Republicans and Democrats collaborating was refreshing to people," Axelrod later said, "even if they didn't like every element of the compromise."

"Winning the Future" just might appeal to those voters craving presidential leadership aimed at transcending partisan bickering to achieve the shared goals of the nation.

—〰—

ON JANUARY 25, 2011, OBAMA ENTERED THE HOUSE OF REPRE-
sentatives. The Tucson shooting shaped the mood. The guests in
the First Lady's box included Daniel Hernandez, an intern for
Giffords who had helped save her life during the attack, and the
parents of Christina Taylor Green. Senators and representatives
were wearing black-and-white ribbons to honor the victims of
that tragedy. And this year, at the suggestion of Senator Mark
Udall, a Colorado Democrat, Republicans and Democrats were
sitting together—not separated into partisan blocs.

Obama began by noting the empty chair of Gabby Giffords.
"What comes of this moment will be determined not by whether
we can sit together tonight, but whether we can work together
tomorrow," he said. He did not mention gun violence—an omis-
sion that disappointed gun control advocates who had hoped for
some reference.

Obama noted that the economy was growing, but pointed
out the American Dream remained under great pressure from
domestic and global forces. He was echoing the speeches he made
during the 2008 campaign, before he (and the rest of the nation)
had been forced to cope with an economic collapse. To win the
future, the president maintained, the nation must "out-innovate,
out-educate, and out-build the rest of the world."

The country did have to erase its deficits, he said, but to be
competitive in the global economy, the government must invest
in information technology, biomedical research, clean energy
technology, and other cutting-edge fields. There would have to
be money spent on education reforms, math and science teachers,
and tax credits for college tuition. Government spending—that
is, investments—was needed to bolster the infrastructure, from
high-speed rail to high-speed Internet, from crumbling roads to
falling-down bridges.

It was a boldly moderate speech. Obama proposed nothing radical, as he expressed an apple-pie vision that in the not-too-distant past would have won much bipartisan support. He did not unfurl a laundry list of legislative measures for the coming year. (With the Republican House, what would be the point?) The president did not dwell on health care. There was no reference to Wall Street. His two paragraphs on the Afghanistan war were obligatory. He wasn't preaching austerity; he was selling American leadership.

The critical question, Obama insisted, was "whether new jobs and industries take root in this country or somewhere else."

He was talking about jobs in the future, not the present. And his message was stark: if we don't get our act together, our clock will be cleaned by the Chinese and others. China, he noted, was "home to the world's largest private solar research facility and the world's fastest computer."

The president was certainly correct about the necessity of prepping for the future. But would embracing the long view persuade voters that he was doing all that was necessary *right now* to kick the sluggish recovery into a higher gear and, thus, deserved their backing in the oncoming battles with the Republicans, as well as the 2012 election? He didn't address the immediate frustration or anger felt by Americans. He was presenting himself as an optimistic and responsible guardian of the economy in contrast to the gripe-and-slash Republicans. But to anyone worried about being unemployed in the weeks and months ahead, neither side was providing much comfort.

Obama did not duck the deficit. He proposed freezing non-defense discretionary spending for five years. He pledged to work with Congress to reduce Medicare and Medicaid costs, the single largest contributor to the long-term deficit—without revealing any details that could become targets, per Plouffe's strategic desires. He called for a "bipartisan solution to strengthen" Social

Security, without offering views as to what should be done. (He did rule out "slashing benefits" or privatizing the program.)

The president was attempting to construct a firewall between decreasing government spending and increasing necessary government investments: "Cutting the deficit by gutting our investments in innovation and education is like lightening an overloaded airplane by removing its engine." And he noted that the nation faced a trade-off: it could extend the Bush tax cuts for the wealthy, or it could fund schools and scholarships.

"We do big things," Obama said, and it would take momentous action in the months ahead to win the future.

The speech did set up the fundamental political battle that would rage until Election Day 2012. Obama wanted to use government to revive the US economy. But he had calculated that he could do so only if Americans believed he was simultaneously tightening the belt of the bloated beast in Washington.

The Republicans would continue to reiterate their mantra: the only thing we have to fear is government spending and debt. In a reply to Obama's speech, Paul Ryan gloomily offered two points: public debt is destroying the United States and limited government is the solution to the nation's woes.

A basic and ideological disagreement was at hand: government is evil, government can help. Obama was conceding part of the argument (yes, we must do something about spending) to win the argument (we must engage in joint action to survive and succeed in the global economy). To win the future—and the next election—the president had to calculate the right ratio.

The "Winning the Future" speech itself was no game changer. John Dickerson of *Slate* observed, "The line sounded more like the title of a self-help seminar, with Obama in the role of Tony Robbins." The deficit hawks sniffed at the address, with the *Washington Post* editorial board huffing that the president had failed "to prepare Americans for fiscal austerity."

Within the White House, according to one aide, the speech "was deemed useful, but not a wild success. . . . It was very Plouffe. It was snickered at by pundits, and he doubled down and stuck to it."

THAT WEEK, THE BUDGET BATTLE PROCEEDED. THE HOUSE REpublicans approved a resolution endorsing a $100 billion cut in domestic spending for the year. No details were included, but such an assault could mean a 20 percent or more reduction in the budget of the Environmental Protection Agency, tremendous cuts in Pell grants that help low-income students attend college, significant reductions in aid money for Afghanistan and Pakistan, and much more. Senate Democrats complained that all this reduced spending could cost up to one million jobs.

About this time, the Congressional Budget Office revised its deficit projection for 2011—upward by $414 billion to $1.48 trillion, which would be close to 10 percent of the gross domestic product. And Treasury was now predicting that the US government would hit its debt ceiling sometime between April 5 and May 31.

Boehner had recently issued a not-too-veiled threat regarding the debt ceiling: "President Obama and congressional Democrats have been on a job-destroying spending spree that has left us with nothing but historic unemployment and the most debt in US history. If they want us to help pay their bills, they are going to have to start cutting up their credit cards."

Boehner was being disingenuous. Raising the debt ceiling had little to do with paying the bills of the Obama administration. The biggest contributors to the debt had been the wars, tax cuts, and policies of the Bush-Cheney administration. But though Boehner and Cantor, in private communications with the White House, had said they did not intend to fool around with the need to raise the debt ceiling—acknowledging that default could be catastrophic—they were willing to demagogue in public.

White House aides adopted the view that it would be best generally to disregard the hyperbole hurled by Boehner and other GOP leaders because that's what they had to do to keep their Tea Party base happy. As long as Boehner seemed reasonable in private, they assumed, that was what mattered. But such rhetoric hardly created a political environment amenable to negotiation and compromise.

To win the future, Obama would have to win the next few months. And that was looking increasingly problematic.

On February 8, the Senate Democrats boarded luxury buses and headed to a posh resort outside Charlottesville, Virginia. There were indoor tennis courts and a spa available to the senators and their spouses. But the lawmakers had left Washington not to relax but to cogitate on the issues they would confront in the coming year. One session that would stick the most with many of them was not led by a policy expert but by Democratic pollster Geoff Garin, who had one major point to impart: you have to be serious about deficit reduction or the voters will not listen to you.

With Sperling sitting in on the presentation, Garin reinforced the White House view that Democrats had to up their game on deficit reduction. His firm had conducted extensive polling and focus groups. He told the senators that voters saw jobs as the most pressing priority. This might seem to support those Democrats who believed Obama had gone too far overboard on the deficit-reduction cruise. But when asked what the president and Congress should do to boost job creation, most voters said reduce the deficit and the debt. They had imbibed the GOP message: the problem with the economy was governmental red ink.

That was not accurate. The financial crash that triggered the economic collapse was unrelated to federal deficits. But Garin measured voter perceptions, not whether voters were correct. And he told the senators that voters would not listen to what the Democrats—including the president—had to say about jobs and investments if they did not sense that the Democrats were willing to wrestle the debt monster to the ground.

But there was good news. Garin had tested three different "frames." One summarized the dominant Republican view advocated by Paul Ryan. In short, it went like this: the main problem is that government has grown too big and is doing too many things that ought to be left to the private sector, and we should not be raising taxes on anyone; we should reduce or eliminate many government programs and return to limited government as envisioned by the founders.

The second was dubbed a "Democrat/Deficit Hawk" frame: the problem of the deficit and debt is real, and to deal with this we will have to engage in shared sacrifice, with everything on the table, including Social Security, Medicare, defense spending, and taxes.

The third frame was labeled "Democrat/Populist": of course, we must cut the waste out of government, but the budget should not be balanced on the backs of the poor and working families; instead Congress ought to stand up to the special interests and cut corporate subsidies and make sure the wealthy pay their fair share of taxes.

Both Democratic frames were supported by almost 70 percent of those Garin polled. The GOP/Ryan frame was backed by 56 percent. And independent voters fancied each of the Democratic frames by 12 to 15 points. (When forced to choose between the two Democratic messages, voters opted for the populist one by a 55 to 35 percent margin.)

This was heartening for Democrats. Their two frames clobbered the Ryan message. But the Democrats were not rallying around either of these frames. They were divided—with Obama trying to straddle the line. And Obama had not yet come across as a leader on that front. "He still seemed to be reacting to the Republicans," a top Democratic strategist familiar with this polling data later said. "People heard what the president was saying as a tactic, not a position."

Voters, Garin informed the senators, were not merely looking for a politician with a certain stance; they wanted commitment to the cause.

Asked whether they would be more worried by a senator who went too far in cutting government programs (including Social Security and Medicare) or by a senator who did not go far enough in reducing the federal deficits, those polled by Garin fretted more about the insufficient slasher, 50 to 40 percent.

Democrats, Garin advised, had to "lean into" the deficit debate. They could not contend the deficit was not of vital importance. Nor could they advocate dealing with it down the road. It was not merely a made-up issue promoted by centrist elites in Washington.

Garin's pitch resonated with the senators. Word of this presentation spread throughout Democratic circles in Washington. It confirmed Sperling's belief that the president and his party could not ignore or escape the deficit politics. To be able to spend—to win the future—they would have to convince voters they were willing to cut.

Chapter Six

LEANING FORWARD

IT WAS A ROUTINE CALL. EVERY FEW MONTHS PRESIDENT OBAMA checked in with Egyptian President Hosni Mubarak. The two would discuss regional matters, the latest on Middle East peace prospects—if there were any—and developments related to the fight against Islamic extremists, the usual. On this day, January 18, 2011, Obama added a warning to the conversation.

A month earlier, Mohamed Bouazizi, a twenty-six-year-old fruit and vegetable peddler in the Tunisian town of Sidi Bouzid, had doused himself with paint thinner and set himself on fire outside a local municipal office. It was a protest. Earlier, the local police had confiscated his cart and beaten him; local officials had refused to hear his complaint.

Bouazizi's act of profound desperation triggered demonstrations in his town against unemployment, police violence, and the lack of human rights. The protests spread and reached the capital of Tunis, threatening the repressive regime of Zine al-Abidine Ben Ali. Police used violence to thwart the protesters, but the demonstrations continued to grow, and in January, Ben Ali was compelled to flee to Saudi Arabia.

The lesson of Tunisia, Obama told Mubarak, was that a strongman leader could not rely on a crackdown to navigate his way through a popular uprising. The best course was to get ahead of popular sentiment by implementing moves toward democracy, transparency, and accountability. What happened in Tunisia could occur anywhere, Obama said.

A week earlier, during a much-noticed speech in Doha, Secretary of State Hillary Clinton had declared that the "region's foundations are sinking in the sand" and that the leaders of the broader Middle East, in response to the demands of their citizens, needed to implement extensive economic, social, and political reforms. Obama was echoing that message: these leaders must respond to criticism with reform, not repression.

Obama was telling Mubarak that he could face a Tunisia-like challenge. Mubarak, though, gave no indication he was concerned—or thinking ahead. The exchange was unsatisfactory for Obama. The call ended without Mubarak expressing any commitment to reform.

LONG BEFORE TUNISIA'S JASMINE REVOLUTION, OBAMA HAD contemplated the possibility of democratic change in North Africa and the Middle East. The previous August, he had sent a memo, known as a Presidential Study Directive, to Joe Biden, Robert Gates, Mike Mullen, Hillary Clinton, and other members of his national security team, requesting they devise strategies for supporting political reform when citizens in this region challenge autocratic leaders allied with Washington.

At that point, Obama was growing frustrated with what appeared to be his limited options for dealing with overseas issues of human rights and democracy. The standard operating procedure was to issue a statement of calibrated condemnation. "He was looking for greater policy tools than just the strength of his rhetoric," recalled Ben Rhodes, the deputy national security adviser for strategic communications, "[and he was] fed up with the status quo—particularly the implicit support for repressive regimes, coupled by the occasional statement of criticism. He wanted to be more strategic about empowering new voices and supporting reform."

The president was being prescient. In his first year in the White House, he had not made the promotion of democracy abroad a prominent priority. During a high-profile speech in Cairo in June 2009, in which Obama tried to reset US relations with the Muslim world, he only gently urged democratic reforms in the region and declared that "no system of government can or should be imposed upon one nation by any other"—a clear reference to President Bush's invasion of Iraq. The following week, when an apparently crooked election in Iran was followed by popular rebellion, Obama did not rush to back the uprising. (His aides maintained that he did not want to intervene in a manner that would afford Tehran the ability to cast the protests as an American-fueled endeavor.)

But Obama was not convinced that the stability in the region was so stable. In his directive, Obama referred to "evidence of growing citizen discontent with the region's regimes" and noted that American allies would probably "opt for repression rather than reform to manage domestic dissent." He was particularly worried about Mubarak, who had ruled as an autocrat for nearly three decades and was preparing to have his son Gamal succeed him, while overseeing sham parliamentary elections.

Getting caught on the wrong side of a revolution—and in bed with a dictator—would be bad for American interests, Obama observed in this memo: "Increased repression could threaten the political and economic stability of some of our allies, leave us with fewer capable, credible partners who can support our regional priorities, and further alienate citizens in the region. Moreover, our regional and international credibility will be undermined if we are seen or perceived to be backing repressive regimes and ignoring the rights and aspirations of citizens."

In response to the directive, a White House team of National Security Council staffers pulled together an interagency review of policy. The group met weekly. The CIA crashed out paper after paper. The participants considered big questions. How could Washington influence events in an autocratic country? What were the risks to US interests of greater openness in the region? And the review zoomed in on Egypt as an archetypal case in which political reform collided with US security interests.

One question this group weighed: Could Obama publicly criticize Mubarak as a means of encouraging political reform, if the United States needed his assistance in dealing with the Israeli-Palestinian conflict and terrorism issues?

Yes, these officials concluded, because Mubarak engaged in these activities due to his own interests, not because he was a pal of Washington.

"Obama saw change was coming," Samantha Power, a National Security Council staffer involved in the review, said. "He didn't identify the precise timing of the change. But he knew we had to be building a foundation for the morning after."

OBAMA WAS TRYING TO RECONCILE TWO FUNDAMENTAL IM-pulses of US foreign policy: realism and idealism. He had campaigned as a liberal realist. He had opposed the Iraq war and talked about prosecuting the Afghanistan war in a more effective manner—not to bring democracy to that nation, but to beat back the threat from al-Qaeda and its allies.

Shortly before announcing his presidential bid, Obama had called for a foreign policy "no longer driven by ideology and politics but one that is based on a realistic assessment of the sobering facts on the ground and our interests in the region." On the campaign trail, he had suggested American troops ought not to be deployed to prevent genocides or humanitarian disasters. "The truth is," he said at one point, "that my foreign policy is actually a return to the traditional bipartisan realistic policy of George Bush's father, of John F. Kennedy, of, in some ways, Ronald Reagan."

But in office Obama sensed that a reckoning was near, and the policy review was close to completion when Tunisia exploded. The assessment had reached several preliminary conclusions: political reforms that fall short of democracy won't lead to stability, democratic transitions succeed when a pact (informal or formal) is struck between the soft-liners of the old regime and the upstart democratic activists, and a push for human rights and democracy in the region would be effective only if integrated throughout the US government. The group had begun drafting a presidential directive that would present a variety of options. But nothing was yet final.

When the Tunisian uprising occurred, Obama let his aides know he was annoyed. "We were caught off guard," a senior administration official recalled. "And he wanted to be sure we didn't get caught off guard by additional unrest elsewhere."

But that's exactly what happened.

ON JANUARY 25, PRO-DEMOCRACY PROTESTS DETONATED IN Egypt, with close to one hundred thousand Egyptians pledging to participate in the demonstrations through a Facebook group. By midday, protesters were in Tahrir Square in Cairo. A couple of deaths were reported. Government authorities blocked Twitter in Egypt, detained and roughed up reporters, and shut down websites of independent newspapers.

The Obama administration's initial response was tepid and

traditional. Secretary Clinton noted that the US government supported "the fundamental right of expression and assembly" and urged restraint, but she said, "The Egyptian government is stable and is looking for ways to respond to the legitimate needs and interests of the Egyptian people."

Just two weeks earlier, Clinton had warned leaders of the region to get right with reform. Now she seemed to be cutting a US ally some slack. She and other administration officials already knew Mubarak was not interested in reform.

"It was all fine and good for the United States to talk about being for reform," a White House staffer later said. "But when it involved a country we had multiple interests in, it was more challenging. Egypt was clearly a test case for the administration's commitment to democracy and human rights."

That night, Obama was due to deliver the annual State of the Union speech. During the drafting of the address, White House aides had debated whether the president ought to refer to the Tunisian revolution. "There was a concern among some that if the president sided too much with the activists in the street, it would cast doubts on our alliances with autocrats like Mubarak in Egypt and [President] Saleh in Yemen," according to a former White House aide.

But the forward-leaning camp won and squeezed a line into the speech. Obama noted that the United States "stands with the people of Tunisia, and supports the democratic aspirations of all people."

He didn't mention Egypt. It was too soon. But the "all people" was a slight nod to those Egyptians taking to the street. Following the speech, Robert Gibbs sent out a statement noting that the Egyptian government should "be responsive to the aspirations of the Egyptian people" and "pursue" political reforms—nothing specific about free and fair elections.

Egypt certainly posed a greater test for the administration

than Tunisia did. Mubarak, an autocratic (and corrupt) thug at home, had long been a useful ally, a crucial interlocutor in the Middle East, who helped preserve an imperfect peace with Israel. He was a partner in quashing al-Qaeda (though his intelligence service had a reputation for using torture and harsh means in this endeavor). Administration officials, such as Biden and Gates, had known him for years; American military officials were close to Egyptian counterparts. And each year, the United States handed Cairo $1.5 billion in aid (as a result of the 1979 peace accords), of which $1.3 billion was used to purchase US military equipment. In essence, Washington gave Egypt a gift card to be used with US military contractors.

Obama and his advisers had to consider competing interests. The president was on record as supporting the advancement of political rights in the Middle East. Could he turn his back on an Egyptian movement for change? Then again, could he dump a longtime partner who had contributed to stability in the region? Moreover, what might replace Mubarak's regime? The out-lawed Muslim Brotherhood or other Islamic extremists? Better the devil you know, some Obama aides argued. And this went beyond Egypt. If Obama tossed Mubarak aside, would other US allies in the region, such as the Saudis, question their working relationships with Washington?

Biden, Clinton, and Gates—no surprise—were part of a stability-first camp. In the past, when Biden met with or spoke to Mubarak, White House aides who wanted the vice president to push the Egyptian leader toward democratic reforms found Biden reluctant to lean too hard on Mubarak. "It almost became a joke with us," a White House aide said. "In one case, Mubarak was back from a surgery, and Biden said he didn't feel comfort-able pressing him."

Clinton, according to a former State Department official, "took a more conservative view at first. She noted it had to be recognized that Mubarak had played a stabilizing role and that had to be taken into account. She agreed there had to be a transition. But sooner or later? Her initial inclination was to give him the opportunity to make a dignified exit. She was reluctant to push him aside."

Obama, one aide recalled, "was good at listening intently and asking probing questions from different directions, and you don't know where he's going to come out. But he figured out early this was moving in a decisive direction, and he needed to be moving in the same direction—while being sensitive that the United States had been a partner of Mubarak. He wanted to give Mubarak good advice to create a soft landing."

Rhodes noted, "The president had a better sense than others of what had to be done. He has an instinct for social movements and nonviolent change movements. He had expected the unrest in the region to spread. He wanted to make sure we embraced it."

But embracing a street revolution aiming to overthrow an ally is not an easy business. First, there would be some tap dancing.

ON THE SECOND DAY OF PROTESTS—AS MUBARAK BANNED public gatherings—Clinton called on the Egyptian government to "exercise restraint" and urged it to allow peaceful protests. She called the moment "an important opportunity" for Mubarak to implement reforms. But she didn't ask Mubarak to allow real elections (or ones without him)—a key demand of the protesters.

About that time, Gibbs, on Air Force One as it carried Obama to Wisconsin, where he would tour local businesses, was asked by a reporter whether he still backed President Mubarak. He answered, "Egypt is a strong ally." It was an ambiguous statement. Was he referring to Mubarak or the nation?

Critics at home and in Egypt wondered aloud if Obama was

moving too hesitantly. Nobel Prize winner Mohamed ElBaradei, a leader of the political opposition in Egypt, blasted Clinton for saying that Mubarak's regime was stable. Later on, Shady el-Ghazaly Harb, another opposition leader, said that protesters interpreted the Obama administration's initial response as a clear but discouraging signal: "Go home. We need this regime."

Obama and his aides were walking a fine line. They supported the demonstrators' democratic aspirations, but they didn't want upheaval in that country. Under the Egyptian constitution, if Mubarak were forced out, the election for a new president had to occur within sixty days. That could lead to the banned but highly organized Muslim Brotherhood ruling the largest Arab nation. Egyptian officials, of course, were telling American officials the demonstrations would soon fade, and they were claiming that the protesters were terrorists. But from the start, Obama was inclined to see the protesters as diverse and largely secular. And whatever course the US government would follow, the president instructed his national security aides to make certain that events in Egypt did not become about the United States. This had to remain a made-in-Egypt affair.

COME JANUARY 27, THE THIRD DAY OF THE PROTESTS, WHITE House aides began telling reporters that the administration was on a dual track, reaching out to the Egyptian activists and pressuring Egyptian officials to heed the demands for reforms. Yet several different messages were on display in public. At the daily press briefing, Gibbs, when asked if Mubarak had Obama's full support, dodged the question, saying, "This isn't a choice between the government and the people of Egypt." (Actually, it sort of was.)

Then Gibbs noted that Mubarak "is a close and important partner," before he commented, "We will continue to push and prod [Mubarak toward] a place where a political dialogue can take place."

He declined to say whether Mubarak's departure would be a positive development, and he stated that the administration was not "taking sides" in Egypt. Still, he stated the White House's "desire to see in Egypt free and fair elections."

At the State Department, spokesperson P. J. Crowley noted that the US government was not telling the Egyptian government it had to institute reforms—but he cited the *need* for reform in Egypt. "We're offering our advice to Egypt," he explained. "But what they do is up to them."

So the administration was pushing and prodding and looking for real elections, but not telling Mubarak what to do. It was complicated.

Yet that same day—as news outlets were reporting that a thousand or so Egyptians had been arrested—Biden, appearing on PBS's *NewsHour,* said it was not time for Mubarak to step aside. Biden saw no connection between Tunisia and Egypt and said he did not consider Mubarak a dictator. (Biden had not prepped much for the interview, according to a senior administration official, and had replied "humanly" because he had known Mubarak for many years.) Meanwhile, State Department officials were sending out tweets, calling on the Egyptian government to stop blocking social-media sites. Administration officials told reporters that Obama was ready to step up his criticism of Mubarak, if Egyptian authorities intensified their crackdown.

This was hardly a unified message. With so many different public signals, some saw Obama on the side of the protesters; others believed he was rooting for Mubarak and didn't want to shove him aside. His national security team remained split.

"Some aides in the White House wanted to take a harder line on Mubarak," according to a senior administration official, "but there was reluctance in State and the Pentagon."

———

GIBBS WAS HAVING A ROUGH TIME BEING A SPOKESPERSON FOR A divided administration—particularly when senior members of Obama's national security team did not want the White House to say anything at all about the uprising.

Clinton and Gates were "wildly cautious," according to a top administration official at the time, "and did not want us to make any comments or remarks that would make it seem we were not sticking up for an ally." In meetings, Bill Daley asked why the president should interject himself into the Egyptian drama. David Plouffe worried that if the president said anything supporting the protests, he could end up owning whatever happened in Egypt.

Obama, though, realized that millions of people, not just in Egypt but throughout the Muslim world, were watching the United States—and him—to see how the Americans would respond to this popular call for democracy.

On most days, Gibbs kept his daily press briefing as mundane as possible; his job was not to make news. But now the whole world was watching every word he uttered for signs of Obama's intentions regarding Mubarak. And Gibbs was practically out the door, scheduled to leave his post as press secretary in two weeks. He saw this as his last big mission at the White House and told Rhodes he was determined not to "fuck it up."

With reporters pressing for Obama's reaction to the uprising, Gibbs believed it was untenable for the president to be mute and for the White House to pretend he didn't have an opinion on the historic events in Egypt. And Gibbs was not alone. In meetings of the national security team, UN ambassador Susan Rice, Samantha Power, and Ben Rhodes argued that Obama had to address the events and show he favored peaceful protests in pursuit of democratic reform. He didn't have to embrace the demonstrators fully, but silence from the president would under-mine his Cairo speech.

"It was hard for them to go up against Gates, Clinton, and Donilon," recalled an official who attended those sessions. "What those guys hate the most is the unknown and uncertainty. And we had a great relationship with the Egyptian government and military."

As some aides saw it, Washington was stuck with Mubarak, like it or not. "There was a certain resignation among several of the principals," a senior administration official recalled, "that it was not possible to get on the right side of history. The history of our relationship with Mubarak doesn't disappear because the president comes out and speaks. And even if your heart was with the people in the street, you still had to ask yourself if it was a good idea to put Barack Obama at the center of Egypt's biggest moment."

ON JANUARY 28—A DAY WHEN TENS OF THOUSANDS OF PROTEST-ers in Cairo fought police (who were using tear gas canisters la-beled MADE IN THE USA) and took control of Tahrir Square—the army was deployed, ElBaradei was placed under house arrest, the Egyptian government shut down Internet access throughout the country, and Mubarak ordered a 6:00 P.M. curfew.

Obama held a meeting with his National Security Council in the Situation Room and told members to see this as an oppor-tunity. He knew Biden, Gates, and Clinton had deep experience with Mubarak and were not eager to cut him loose.

We can, the president said, look at this as a huge problem. But it's also a chance to align US values with its actions and pursue a different model of stability. There was no going back to the old ways.

"This was not the view of the whole government," a participant in the meeting later said. "For much of the government, Egypt was stability."

The president "was consistently the most forward leaning," Rhodes subsequently maintained.

The United States ratcheted up its response. Gibbs and Clinton publicly demanded that Egypt restore Internet communications. At the daily press briefing, Gibbs said Mubarak had to address the "legitimate grievances" of the Egyptian people "immediately" with "concrete reforms." And he issued a threat: the administration would be reviewing US assistance to Egypt "based on events that take place in the coming days." (Two days later, Clinton would say, "There is no discussion as of this time about cutting off any aid"—yet again sending a mixed message.)

Following this daily briefing, Gibbs was talking with Obama in the Outer Oval Office. You have to say something, he advised the president; these protests were historic. Just then, Denis McDonough, the deputy national security adviser, came by and informed the president that the national security team, which had just been meeting in the Sit Room, was recommending that he still not comment on the uprising. Gibbs thought that was crazy and said so.

The president asked if his national security advisers were still in the Situation Room. They were—and Obama headed downstairs. There was a brief discussion. Obama indicated he was predisposed to issuing a statement.

"By the time he left," one participant said, "everyone was in favor of him saying something."

Soon after, when it was past midnight in Egypt, Mubarak appeared on state television and said he'd requested that the government resign so he could appoint a new one. He said nothing about resigning himself.

Obama, joined by Gibbs, McDonough, and Blinken, watched the speech on a small television screen in the Outer Oval Office.

Afterward, White House aides debated how Obama could pressure the Egyptian leader, who clearly believed he could ride out the protests.

Obama called Mubarak. He didn't encourage him to leave office. But he told the eighty-two-year-old autocrat that he had to implement extensive reforms quickly. Mubarak appeared to be in denial about the seriousness of the protesters' desires and demands. The demonstrations were a result of outside interference, he said. Obama wasn't getting through.

Following the call, Obama went on television to issue his first public comment on the protests. The Egyptian people, he said, had the right to protest and "to determine their own destiny," and the United States would "stand up" for the rights of the Egyptian people. He called on Egypt to lift its restrictions on communications and social networks, and he reported that he had just told Mubarak to "take concrete steps" to address the grievances of the Egyptian people.

Obama did not signal that Mubarak had to resign. In that regard, he was not in sync with the protesters in Tahrir Square. Obama and his aides worried that if he took that ultimate step, Washington would lose whatever leverage it had with Mubarak. But nevertheless he had made a forceful statement of support for Mubarak's foes in the streets.

THE WHITE HOUSE, THOUGH, REMAINED IN CAREFUL CALIBRATION mode. The next day, protesters defied Mubarak's curfew and flooded Cairo's streets, as Mubarak appointed intelligence chief Omar Suleiman (who had long worked with Washington while overseeing a spy service engaged in abusive practices) as his vice president and took steps to form a new government.

Crowley, without consulting other officials, tweeted that

Mubarak's moves were insufficient: "The Egyptian government can't reshuffle the deck chairs and then stand pat."

It was a blunt assessment, not out of line with the administration policy, but White House aides were angry that Crowley was pushing the envelope on his own. E-mails flew between 1600 Pennsylvania Avenue and Foggy Bottom, and a new rule was imposed: all future Crowley tweets would have to be cleared.

"This little episode highlighted the difficulty we faced," said a senior administration official. "How much did we have to say publicly to maintain credibility with the protesters, and how much did we have to hold back in order to be able to effectively communicate privately with Mubarak and his allies."

At the White House, Obama's top aides—Biden, Daley, Clinton, CIA chief Leon Panetta, and others—huddled in the Situation Room for a meeting, led by Donilon, and they agreed it was time to start easing Mubarak out—somehow.

Clinton suggested sending Frank Wisner, a former diplomat, to deliver the message to the Egyptian strongman. Wisner had served as US ambassador to Egypt in the Reagan and first Bush administrations, and he remained a close pal of Mubarak. If anyone could politely encourage Mubarak to leave, it was Wisner.

In the afternoon, Obama convened another meeting of his national security team, with Plouffe and Gibbs also attending, and afterward the White House put out a press release reporting that Obama had "reiterated our focus on opposing violence and calling for restraint; supporting universal rights; and supporting concrete steps that advance political reform within Egypt."

It said nothing about the president trying to persuade Mubarak his time had come. In the streets of Cairo, protesters, unfamiliar with the behind-the-scenes actions, denounced Obama for not supporting their full demands.

Some of the stability-first crowd within the administration

were still waiting to see—or hoping—that Mubarak might hang on. Obama, however, thought reform was inevitable in Egypt. That's why he had commissioned that review the previous summer. So at least he had prepared himself for the moment at hand.

"If you already have gotten to the thought that Egypt would not have a dictator forever," Samantha Power recalled, "now you're moving to the business of managing not only the present but that very different future, and every day you are laying the foundation for the relationship you will have with an as-yet-unknown successor."

ON JANUARY 30, WHILE THE PROTESTS CONTINUED AND DEMON-strators in Tahrir Square vowed not to leave until Mubarak resigned, ElBaradei scoffed at Obama's public position: "To ask a dictator to implement democratic measures after thirty years in power is an oxymoron."

The Egyptian uprising was demanding much of the White House's attention. In meetings with the president, Plouffe advised him not to get sidetracked from his economic message. But it was difficult for anyone, including the president, not to be fixated upon the Egyptian drama.

As Obama was managing a divorce from Mubarak, the president's political opponents were split. Speaker John Boehner noted that Obama "so far has handled this tense situation pretty well"—a rare display of bipartisanship. But not all of Boehner's ideological comrades agreed. One band of conservatives—particularly neoconservatives—decried Obama for moving too slowly to side with the Egyptian protesters. Other right-wingers groused that Obama was kicking Mubarak aside and paving the way for Islamic extremists to seize control of the country.

On Fox News, Glenn Beck depicted the Egyptian uprising

as a part of a global plot engineered by both communists and Islamic extremists. Republican Senator Mark Kirk of Illinois tweeted, "The U.S. should continue aid to Egypt, especially military. The government should not collapse, giving radical Muslim Brotherhood an opening."

On another front, the Saudi government was using its multiple contacts with the Obama administration to deliver a desperate message: don't abandon Mubarak. The Saudis feared contagion.

THROUGHOUT THE FIRST WEEK OF THE UPRISING, GATES AND Mullen were warning their counterparts in the Egyptian military not to intervene in any violent manner. For years, democracy and human rights advocates had criticized the United States' close relationship with the Egyptian military; here was the silver lining. And on January 31, as Mubarak claimed he was amenable to negotiating with opposition parties, the Egyptian military declared it would not use force against the protesters.

Wisner landed in Cairo that day, and the Obama administration was coy about his role. Crowley denied Wisner was an envoy and maintained he was visiting Mubarak merely as "a private citizen"—who would "reinforce" the "very clear message" the administration had already sent Mubarak. (The next day, the State Department would clarify that the administration had indeed asked Wisner to undertake this trip.)

The message, however, remained unclear. Media accounts were full of quotes from anonymous administration officials saying that now the White House desired Mubarak's departure but was unsure how firmly to push him, for there was still a chance he could gut it out. But when a reporter asked Gibbs why the president was not taking a public position on whether Mubarak should stay in power, Gibbs replied, "That's not for us to determine."

Gibbs reiterated Clinton's call for an "orderly transition" in Egypt—without stating specifically what that would entail. Pressed repeatedly on whether Obama was urging a change in the Egyptian government, Gibbs said, "No, we're calling for a change in the way the country works." Asked whether Egyptian demonstrators could "infer a subtle support for Mubarak" from Obama, Gibbs didn't say yes; he didn't say no.

Months later, Gibbs observed, "The hardest-line protester thought we were on the side of an autocrat. And the autocrats in the region thought we were on the side of the most ardent protester. And if Obama came out and sided firmly with the protesters, the Egyptian government would have used that against them."

WITH EVENTS INTENSIFYING IN EGYPT—ON FEBRUARY 1, TAHRIR Square was packed with protesters, and more than a million Egyptians attended antigovernment demonstrations across the country—it was becoming tough for the Obama administration to sustain its current posture. It was calling for change without taking a public position on the fundamental question posed by the uprising: Should Mubarak stay or go?

In Cairo, Wisner met with his old friend and encouraged Mubarak to say he would not run for reelection, meaning that his reign of power would soon end. (Whether Wisner actually told Mubarak that he ought to consider resigning sooner was unclear.)

"After that first conversation with Wisner, it became obvious Mubarak was not interested in real change," a former senior administration official later said. "Our assessment was that people near Mubarak were not telling him what he needed to hear—or he was not listening. He thought he could manage it."

That afternoon, the principals of the National Security

Council were again in the Situation Room—but without Obama. The question was whether to recommend to the president that he call Mubarak or issue a statement to increase pressure on the Egyptian leader. Susan Rice was advocating that the administration strike a harder stance on Mubarak. Clinton was hesitant. Obama heard about the meeting as it was under way, and he came in and assumed control.

As he and his advisers discussed their next step, Mubarak delivered a meandering and confusing speech, announcing he would not run for reelection. But he did not say he would resign. In Tahrir Square, the demonstrators erupted in fury. Obama and his aides watched a cable news broadcast of the speech, and when the Egyptian leader was done, Obama said Mubarak's stance was unacceptable; Mubarak could not put the protests back in the box with half measures.

The president was losing patience with Mubarak, and he was irritated that his administration's policymaking apparatus was not maintaining pace with the fast-changing developments. ("We were spending at least half our time keeping up with what was going on," a senior administration official later said, "rather than finding a way to influence unfolding events.")

The Pentagon and the State Department, according to a participant in the meeting, had their "institutional habits" when it came to supporting Mubarak. But, as Rhodes noted, "Obama didn't know Mubarak. He didn't have a long-standing relationship with Mubarak. And it was hard not to identify with people on the street, using Facebook and technology to get connected. That had been our campaign."

After some discussion, Obama said, "I'm going upstairs to call Mubarak and tell him he should step down."

The meeting was over.

In the Oval Office—with Gibbs pacing about, Donilon, McDonough, and Rhodes huddling in one corner, and Daley on the couch clutching rolled-up papers—Obama, at his desk, phoned Mubarak. It was a difficult conversation. Obama in several ways suggested that Mubarak ought to resign. Not running for reelection later in the year was not sufficient. He gently raised scenarios for a stable transition.

An angry Mubarak pushed back. He once again claimed that the Muslim Brotherhood would gain power in Egypt if he were to resign. He repeatedly insisted that the situation in Egypt would improve. That's not our analysis, Obama told him.

"I respect my elders," Obama said. "And you have been in politics for a very long time, Mr. President. But there are moments in history when just because things were the same way in the past doesn't mean they will be that way in the future."

The call ended inconclusively. The two leaders ran out of things to say. Mubarak was set on holding on. The nudging hadn't worked. Now Obama would have to give Mubarak a push—in public. Not a big push, but a push nonetheless.

Obama and his aides quickly drafted a statement. Then he walked to a podium set up in the Grand Foyer, the primary and formal entrance to the White House. In a live broadcast, Obama noted that he had just told Mubarak it was "clear [that] an orderly transition [in Egypt] must begin now."

Now—that was the key word. This was a polite way of saying that Mubarak had to go right away. This was the moment Obama cut the cord.

Protesters and others, however, regarded the statement as a step short of a call for resignation. But Egyptian officials considered it an unambiguous elbowing. They were angry. (The Israelis and the Saudis were also upset.)

The next day, Gibbs, referring to Obama's statement, declared that "now means yesterday."

Privately, Obama seethed as he watched the latest images of violent clashes between protesters and Mubarak's supporters in

Tahrir Square. Clinton called Suleiman, and Gates and Mullen contacted their respective counterparts, all to reiterate Obama's message: a transition to a post-Mubarak government had to start immediately. (European leaders, including British Prime Minister David Cameron and French President Nicolas Sarkozy, followed Obama's lead, demanding that a transition should start "now.")

Obama's effort to increase the pressure on Mubarak drew a variety of reviews. "Obama did not go far enough," a *Washington Post* editorial complained. *New York Times* columnist Nicholas Kristof, who was in Cairo, urged the president to condemn the latest repression more strongly: "You owe the democracy protesters being attacked here, and our own history and values, a much more forceful statement deploring this crackdown." Former Secretary of State James Baker, however, observed, "I think they've done pretty good after some faltering steps to begin with." Obama was still trying to shape events without being too heavy-handed. He was leaning on Mubarak, not kicking him. This approach afforded critics on all sides room for sniping.

THEN FRANK WISNER BECAME A PROBLEM. SPEAKING AT A Munich conference via video on February 5, Wisner, who had left Egypt a few days earlier, said, "Mubarak remains utterly critical in the days ahead as we sort our way toward the future [and he] must stay in office in order to steer those changes through."

Wisner also asserted that calls from outside Egypt for Mubarak to yield power could create a "negative force" within Egypt. Obama's unofficial envoy was bolstering Mubarak's position. Administration officials rushed to say that Wisner was speaking only for himself, not the president. Yet, once again, the administration was seen as sending out mixed messages.

Obama was meeting with Ben Rhodes in the White House when he heard about Wisner's comments. The president was livid. For the past week, he had tried to bridge all the various equities and move deliberately toward a clear signal, realizing this course would upset allies in the region and anger the protesters. Now all that work was shot. Wisner's remarks made it appear that the administration was retreating and backing Mubarak. Obama shooed everyone out of the Oval Office and called Clinton directly.

In response to the Wisner debacle, Obama issued a directive to his aides: the White House had to be in charge of the messaging. He told his staffers he wanted them to be direct when they released public summaries of Obama's and other officials' conversations with leaders in Egypt and elsewhere. There could be no more of the usual candid-exchange-of-views goobledygook. He demanded to be provided daily updates on what officials throughout his administration were saying publicly about events in Egypt. No more conflicting messages, he ordered.

"He got very hands-on and tactical about what we were saying," a senior administration official recalled.

OBAMA WANTED TO DRIVE A HARDER LINE. BUT THIS WAS NOT so easy to do in public if administration officials were looking (or hoping) for a smooth, orderly, and meaningful transition arranged by Suleiman, who was now negotiating with the opposition. Opposition leaders, though, were deriding these talks as political theater on the government's part.

On February 7, as Egypt was entering its third week of protests—demonstrators remained in Tahrir Square, vowing not to budge until Mubarak vacated the presidency—Obama declared that Egypt was "making progress" via the ongoing negotiations, which could be read as an endorsement of a process opposed by many protesters. By issuing a positive comment about Suleiman's efforts, Obama risked being identified with the ruling elite—and an elite that didn't seem that anxious to proceed with reforms. Suleiman was publicly saying that he didn't believe Egypt was ready for democracy, that Mubarak ought not to resign before his term was to end in September, and that it was not quite time to lift the emergency law banning certain political parties.

Still, at the daily press briefing, Gibbs talked up the Suleiman-led negotiations. "We are strongly encouraging the process," he said.

The administration did ramp up its pressure on Suleiman. Vice President Biden called him, and the White House released an unusually detailed and blunt summation of the conversation. It noted that the vice president had urged Suleiman to take specific steps: rescind the emergency law, cease the assaults on journalists and activists, and partner up with the opposition to "jointly develop" a road map and timetable for transition.

"Biden was testing Suleiman to see if he was serious about a transition to democracy," a senior administration official later said.

At the same time, Obama directed the Pentagon at all levels to "flood the zone"—that is, anyone in the Pentagon who had any working relationship with an Egyptian official or military officer was to contact that person and pass along a simple message: no crackdown.

"At the highest levels, we were telling them, If you don't do more, it will be too late," a senior administration official explained. "They would complain to us. We'd say, 'We didn't bring one million people to Tahrir Square. This is about you.'"

THE CROWDS KEPT GROWING IN CAIRO. MILLIONS WERE DEMonstrating throughout the country. And they were demanding that Mubarak and his cronies give up power immediately.

And some American legislators—Democrats and Republicans—were becoming restive. At a hearing of the House Foreign Affairs Committee, Republican Representative Ileana Ros-Lehtinen said the White House had been too pro-Mubarak. Democratic Representative Gary Ackerman squawked, "The Obama administration now appears to be wavering about whether America really backs the demands of the Egyptian people or just wants to return to stability, which is a façade."

The *Washington Post,* noting that US officials had recently highlighted the risks posed by quick elections, was reporting that the Obama administration had decided "to continue backing Mubarak as he clings to power." That was not true. Yet Obama's decision to follow a dual-track strategy—push Mubarak in private, promote the negotiations in public—was wearing thin.

ON FEBRUARY 10, THE WHITE HOUSE FINALLY RECEIVED ENcouraging news. The CIA had solid intelligence that the military would be relieving Mubarak of his primary powers, and a final resolution was being cooked up in Cairo. "We knew this from his inner circle," a top administration official later said.

That morning, CIA chief Leon Panetta, testifying on Capitol Hill, disclosed that he had "received reports" that Mubarak had decided to end his presidency. Other signs indicated Mubarak's self-ouster was at hand. Egyptian officials were telling reporters the end would come that night. An Egyptian military commander in Cairo told protesters, "All your demands will be met today." Wael Ghonim, a Google executive and protest leader, tweeted, "Mission accomplished. Thanks to all the brave young Egyptians. #Jan25."

Next came a surprise. In a speech, Mubarak announced he would delegate authority to Suleiman, but he would not resign. "It's not about me," he said.

The hundreds of thousands of protesters in Tahrir Square, who had been poised to celebrate, were bewildered and enraged. To them, it was about Mubarak.

"Leave, leave," they shouted.

After Mubarak was done, Suleiman ominously warned the protesters to go home.

Obama watched Mubarak's speech on Air Force One, while returning from an event in Michigan. Once he was back at the White House, he gathered his national security team. With Daley, Donilon, Plouffe, and Rhodes in the Oval Office, Obama worked on a statement.

The administration looked dumb. Panetta had said Mubarak would be signing off. "We had been told by a reliable source that Mubarak was going to go," a senior administration official recalled. "But Mubarak changed his mind. He called an audible and surprised us and his inner circle."

The White House released a statement that questioned Mubarak's announcement and his sincerity: "The Egyptian people have been told that there was a transition of authority, but it is not yet clear that this transition is immediate, meaningful or sufficient."

Obama was still trying to make it clear he was aligned with the protesters' aspirations. But as his administration scurried to sort out what had happened in Cairo, he was not yet willing to give Mubarak a public heave-ho.

THE NEXT DAY—SEVENTEEN DAYS AFTER THE PRO-DEMOCRACY protests began—Obama was in a meeting in the Oval Office when a note was handed to him: Suleiman had just announced Mubarak was transferring all his authority to a council of military leaders and had resigned. The United States had been given no heads-up by the Egyptians.

The policy had worked—in that Mubarak, after much behind-the-scenes cajoling from Obama and others, had decided to withdraw peacefully and yield the power he had held firmly for decades. Though Obama had not once publicly called for Mubarak to depart and his administration had put out conflicting (or easy to misinterpret) signals, the president could claim credit for having managed a historic transition that had transpired without a great amount of violence. (About three hundred people were killed in the Egyptian protests, according to Human Rights Watch.)

Obama had not paid heed to his critics and had navigated the competing interests within his own government to facilitate change attuned to his own values and predilections. From the outside, it may have looked sloppy, but Obama felt good and began to work with aides on a statement.

"He was happy about it, energized," Rhodes recalled. "That feeling was not shared by many in Washington."

With Cairo enwrapped in celebration—"The country has been liberated," ElBaradei proclaimed—Obama issued a televised statement.

"There are very few moments in our lives where we have the privilege to witness history taking place," he said. "This is one of those moments."

He praised the Egyptian military for having decided to "not fire bullets at the people they were sworn to protect" and called on it to protect the Egyptians' rights during the transition. He hailed the protesters for causing the "wheel of history" to turn "at a blinding pace." By achieving foundational change through peaceful means, the Egyptian demonstrators, Obama noted, had created a counter to the call of violent extremism: "In Egypt, it was the moral force of nonviolence—not terrorism, not mindless killing—but nonviolence, moral force that bent the arc of history toward justice once more."

He barely referred to Mubarak. Obama didn't once mention his own government's actions.

As he spoke in optimistic terms, administration officials were telling reporters that US strategy in the Middle East would be upended by the uprising—and Israeli and Saudi officials were complaining that Obama had dumped Mubarak before ensuring that Egypt's revolution would not be hijacked by the Muslim Brotherhood or other extremists. Officials in both Israel and the United States worried about the impact on the peace accord between Egypt and Israel. An analysis circulating in the White House pegged the regimes in Yemen and Syria as the next possible targets of internal protests.

Twenty minutes after delivering his statement, Obama escorted Gibbs into the press briefing room. This was Gibbs's last day in the White House. The two men were in a jocular mood.

"Obviously, Gibbs's departure is not the biggest [story] today," the president joshed.

He then presented Gibbs a going-away present: a framed necktie. Minutes before Obama's historic and career-making speech at the 2004 Democratic Convention, his advisers had been locked in a fierce debate over what tie he should wear. Someone suggested the tie Gibbs had on. Obama grabbed it—and somehow it had stayed in his possession until this day.

After Obama left, Gibbs proceeded with his final briefing. Most of the questions concerned Egypt. As usual, Gibbs did his best not to make any news. There was plenty already. But his remarks did indicate that the past two and a half weeks had been difficult for the president and his advisers, as they had been forced to sort out competing impulses and interests.

"The phrase we've used a lot around here is 'threading the needle,'" Gibbs noted, adding, "It was a challenging topic for us to discuss."

That had certainly shown.

GIVEN THAT OBAMA'S DIPLOMACY OF QUIET PRESSURE HAD CRE-
ated a gap between private actions and public utterances, he was
not in the best position for a victory lap. His opponents could
still criticize his performance.

In a speech at a conference of conservative activists, Mitt
Romney, the former Massachusetts governor seeking the Republican presidential nomination, charged that "an uncertain world
has been made more dangerous by the lack of clear direction
from a weak president."

Fox News host Sean Hannity claimed that "we have weak-
ened America's influence in the world" because of "confusion,
incompetence, you know, lack of decisiveness."

More serious thinkers credited Obama with a judiciousness
that turned out to be effective. Robert Kagan, a conservative
scholar, observed that Obama "has done better than his government has done." Nicholas Kristof noted that he had "been critical
of the Obama administration's wishy-washy, weather vane approach to people power in Tunisia and Egypt," but he cheered
Obama's final moves regarding the uprising.

Marc Lynch, a noted Middle East expert, praised Obama's
approach: "For those keeping score in the 'peacefully removing
Arab dictators' game, it's now Obama 2, Bush 0."

Obama's no-drama and patient style had not been satisfying
to those who were inspired by the Egyptian protesters and desired more boldness from the president. But Obama had reaped
the dividends he sought: change without upheaval (that is, no
immediate upheaval).

After Obama received word that Mubarak had resigned, he
called his national security team to the Situation Room. With
Donilon, Clinton, James Clapper (the director of national intelligence), Panetta, McDonough, Gates, Mullen, Daley, John

Brennan (Obama's chief terrorism adviser), and others sitting at the long conference table, the mood was glum.

"Like a funeral," one participant recalled. A strategic partner in an always vexing part of the world was gone. An oasis of certainty in a region of uncertainty was no more.

"Guys, this is a good thing," Obama told them.

He urged them to feel positive about what had just happened. Egyptian political leaders were already declaring their intention to run for president. Obama wasn't a Pollyanna about the future. He knew the Muslim Brotherhood was better organized than the secular opposition, that Mubarak's autocratic party could repackage itself, and that women and minorities could face new challenges.

There was a broader point. Obama told his national security team they needed to engage in deep thinking about how the United States would approach the region after Egypt. The old assumptions would have to be placed aside; the status quo was not firm. He pressed his advisers to devise means for supporting democratic movements over the long term. (This would prompt a White House review of previous transitions to democracy in Indonesia, the Philippines, and Latin America to determine what conditions led to successful change.) Clearly, Obama said, we have to manage whatever comes next in this Arab Spring. This was a story, he predicted, that would continue for years to come.

In three days, antigovernment protesters would be clashing with security forces in Bahrain. In five days, demonstrators would be battling with police in the Libyan city of Benghazi.

Chapter Seven

THE BATTLE BEGINS

JACK LEW OFFICIALLY OPENED THE BUDGET SEASON ON VALENtine's Day 2011, throwing out the first ball—Obama's proposed budget for 2012—with a press conference in the rather plain auditorium of the Eisenhower Executive Office Building.

The fifty-five-year-old Lew, a blend of lawyer, accountant, and policy wonk, had been the budget chief years earlier during the Clinton administration. In the first years of the Obama administration, he had served as a deputy secretary of state. When Peter Orszag vacated the OMB post the previous summer, Obama asked Lew, who began his career as a policy adviser three decades earlier for then House Speaker Tip O'Neill, to return to his old job. Lew, a measured and even-tempered fellow, was known as a sharp negotiator who had acquitted himself well during the Clinton-era dogfights when the Democrats had beat back the Republican revolutionaries.

Now Lew was back—but under much different circumstances. His boss this time was a more careful fighter who did not relish political brawling the way Clinton had, and the current president inhabited a different, perhaps weaker position than

Clinton had held. The economy was in trouble (not expanding as it had during Clinton's presidency), and the Republican opposition was more rigid. Moreover, the stakes were higher. A government shutdown or a national default—outcomes that some Tea Party GOP lawmakers were actually rooting for—could send the economy into a double-dip recession. This budget showdown would not be a replay of the 1990s face-off.

An accomplished numbers cruncher, Lew, the son of a Polish immigrant, tried to see the world beyond the green eyeshade. He once told an interviewer, "Budgets aren't books of numbers. They're a tapestry, the fabric, of what we believe. The numbers tell a story, a self-portrait of what we are as a country."

Lew was not telling a glorious story this Monday afternoon. Nor was it the full story. Obama and his aides were releasing a placeholder budget, hoping to draw the Republicans into a you-go-first trap—that is, as much of one as David Plouffe could manufacture.

The budget Lew described called for $1.1 trillion in deficit reduction over the next decade. Two-thirds of that would come from spending cuts, including a five-year freeze on nondefense discretionary spending—the financing of medical research, food safety, financial regulation, national parks, and all the other programs Congress must fund every year.

Suspending the routine increases (that cover inflation and population growth) in this 12 percent chunk of the overall federal budget would amount to $400 billion in savings. This would bring spending in this category (as a percentage of the gross domestic product) to its lowest level since the days when Dwight Eisenhower was in the White House. Under this plan, over half of federal agencies would experience dramatic reductions. But these were, supposedly, desperate times. The administration predicted a record-setting $1.6 trillion deficit for the year, yet projected it would decline to $600 billion by 2018.

Standing before the room of reporters, Lew identified several programs Obama had okayed for the chopping block: community development block grants, the Great Lakes restoration initiative, airport grants, water treatment projects. The list also included low-income heating assistance—a heartbreaker for Lew, who had helped to develop the program decades earlier. (The administration's proposal would reduce the EPA's budget by 13 percent and eliminate the cap-and-trade plan for reducing climate change emissions.) Almost $80 billion would be sliced out of the Pentagon's purse over five years. Thankfully, the cost of the Iraq and Afghanistan wars would decline in the years ahead.

And Obama's budget would hike taxes on the well-to-do by reducing the value of itemized deductions and eliminating tax breaks for oil, gas, and coal companies. Under this proposal—which was indeed a wish list—the Bush tax cuts for the wealthy would end in 2013.

The good news, Lew said, was that the budget included critical win-the-future investments in education, innovation, and infrastructure, including $148 billion for R&D and $50 billion for transportation projects. Education spending would receive an 11 percent boost.

The first question from the pack of Washington journalists was obvious: Why hadn't the president incorporated the Bowles-Simpson recommendations into his budget?

The commission Obama had created as political cover during the previous year's budget season had set a magic number of $4 trillion in deficit reduction for Washington's debt watchers. Obama's budget weighed in at one-quarter of that amount; it did not propose the significant cuts or savings in Medicare, Medicaid, or Social Security that deficit hawks always desired. The president and his advisers knew they would be assailed by the Republicans and deficit-reduction obsessives for falling so short of the Bowles-Simpson number. But they were sticking to Plouffe's plan, which called for letting the other side go first.

Lew noted that the president would be delighted to work with Republicans on a "long-term solution" to the nation's debt woes. He declined to specify just how such a process should—or would—begin.

One reporter, noting the budget reflected the "winning the future" theme of the State of the Union, asked what Lew would say to an unemployed American who wondered what "short-term, jobs-creation benefits" were contained in the plan.

Lew briefly referred to the unemployment insurance benefits previously extended in the lame-duck tax-cut accord. He then cited the infrastructure boost in this budget. But he didn't have a longer list. Lew said this budget was "designed to deal not just with the short-term, but over the horizon to make sure we have an economy that does provide those opportunities." There was no jobs-jobs-jobs message.

LEW WAS PLAYING OBAMA'S OPENING GAMBIT. IT WOULDN'T SAT-isfy the deficit-fixated pundits and pols of Washington or the "markets." Nor would it hearten progressive advocates or Dem-ocratic allies on the Hill. But David Plouffe and the president were sticking to the script: let Congressman Ryan have a few good weeks as the town's number one deficit slayer and wait for the Tea Party–controlled Republicans to release their long-term slash-slash-slash budget plan before getting serious.

At White House meetings, Plouffe would say, "We're just going to have to endure this."

Obama, Plouffe, and the rest knew that the headlines weren't going to be good—and might not be for a while, with Obama holding back, playing rope-a-dope. And the inevitable criticism and attacks rained down. The *Washington Post* reported, "Obama's budget makes clear that he will not take the lead in that debate: It contains no specific recommendations for tax or entitlement reform."

Representative Paul Ryan exclaimed, "What we have here is a total abdication of leadership and talking points based on gimmicks and cooking the books."

Democratic Senator Kent Conrad, the chair of the Senate Budget Committee, griped, "If we're going to get this debt down to a level that's sustainable, then we've got to do substantially more than $1 trillion worth of deficit reduction in the next decade. We just do."

Obama was allowing himself to be excoriated. *He's no leader. He has no plan . . . and no guts. He's weak.* He was taking the hit—providing Republicans with space for their own extremism. And it seemed to be working.

Responding to Obama's 2012 budget, Boehner vowed that the Republicans would address Medicare and Social Security when they released their budget in the spring.

OBAMA'S 2012 BUDGET WAS LANDING IN A BATTLEFIELD WHERE hot and heavy fiscal policy warfare was already under way. In a few weeks, on March 4, the legislation approved in December to fund the government temporarily would expire. Without another continuing resolution—CR, in Washington parlance—for the rest of the 2011 fiscal year, the government would shut down. The Tea Party Republicans had a hostage situation to exploit.

The Republicans had seized the House on a promise of ripping $100 billion out of the 2011 budget. That was a tremendous sum, especially because they were only targeting nondefense discretionary spending. Soon after settling into their new offices,

House Republican leaders proposed $32 billion in cuts within a CR funding the government through September 30. But Tea Partiers and veteran conservatives inside and outside the House protested that this was not good enough.

"This was the first notice of the power of the freshman Tea Partiers," a House Democratic staffer observed. "They were telling the speaker and the appropriations chairman to go screw yourself."

Three days before Obama unveiled his 2012 budget proposal, the House Republicans, responding to the Tea Party revolt, released a new and expanded hit list of proposed immediate cuts totaling about $61 billion. Though some Tea Partiers still moaned about falling short of the $100 billion pledge, House Republican leaders explained that the CR covered only seven months.

Blood was in the water. Pell grants would be reduced by 15 percent, imperiling college for millions of low-income students. Head Start would be slashed, meaning hundreds of thousands of children from poor families would lose their places in this antipoverty preschool program. The federal food inspection program would be downsized. Funding for schools with needy children would be eviscerated. AmeriCorps and the Corporation for Public Broadcasting would be zeroed out. The WIC program providing nutritional support to low-income women and infants was slated for a $747 million slam. The EPA's budget would be cut by almost a third, with its climate change programs reduced by $100 million. The Securities and Exchange Commission, despite (or because of) its new responsibilities to enforce Obama's Wall Street reform law, was pegged for a whack. Foreign aid would be severely shrunk.

This was a declaration of war. The raise-no-taxes Republicans were raiding dozens of critical programs that provided assistance to low-income families and others. The Senate Democrats released a report detailing the potential damage: 3,000 police officers dismissed, 65,000 teachers fired, 218,000 kids tossed out of Head Start.

"If we follow the House Republicans' road map," Senator Chuck Schumer, a New York Democrat, exclaimed, "it will lead the country over the cliff."

When asked about the Republican plan to hack $61 billion out of this year's spending, Obama responded in a subdued manner, noting he opposed "steps" that would "prompt thousands of layoffs in state or local government" or impede "core vital functions of government," for that would "have a dampening impact on our recovery." He didn't denounce the Republicans for placing the nation on the road to ruin.

At one of the White House's daily press briefings, a reporter noted that Obama's response had been "muted" and asked Jay Carney, the new press secretary, whether the president believed that "the Republican cuts are dangerous . . . going after things like the SEC and the climate change programs."

Carney offered a dispassionate response: "The president has made clear that he doesn't—we cannot support arbitrary or irresponsible or deep cuts." It sounded like routine cant. Carney didn't even bother to mention that the Office of Management and Budget had quietly issued a statement noting that the president would veto the House Republican spending bill if it contained these cuts.

There was a reason for the low volume. Obama and Plouffe were hoping the White House could duck a major clash over the 2011 spending bill. The president and his aides wanted to skate past this quickly, avoiding a government shutdown they feared could screw up the already weak economic recovery. Plouffe saw bigger battles ahead with the 2012 budget and the debt ceiling. The 2011 budget was merely a leftover skirmish best disposed of sooner rather than later. There was no need to rush to the ramparts.

So the message went out from the White House to the cabinet secretaries: your agencies may be targeted for deep cuts, but we're going to take care of it without turning this into an all-out partisan slugfest. A similar message was sent to various policy advocacy outfits, including environmental groups, public health associations, advocates for low-income programs, and the unions: the president won't be fully assailing these spending cuts, but feel free to attack on your own.

White House aides told their progressive pals they were keeping the president from looking overly partisan at the start of this face-off with the Republicans. The liberal advocates pointed out that programs for children and seniors would get clobbered. Moreover, after screaming for months, "Where are the jobs?" the Republicans were now bent on cutting spending no matter the impact on employment. (Mark Zandi, Moody Analytics' chief economist, estimated the bill would cause the loss of seven hundred thousand jobs.) White House aides promised they would do all they could to stem the damage, and explained that in order to be effective Obama had to stay above the fray.

"I kept hearing this above-the-fray line, and I wanted to throw up," a progressive policy advocate said.

ON FEBRUARY 19, BOEHNER AND HIS HOUSE REPUBLICANS, ON A party-line vote, passed the CR, which also contained dozens of provisions known as riders aimed at preventing the president from carrying out various environmental and health care policies that offended conservative sensibilities. Asked about the measure, Carney offered a rather calm response: "We continue to believe that we will be able to work out common ground on these issues."

This disengagement was puzzling to outsiders. A former Obama White House official said at the time, "What the hell are they doing? Why are they not attacking back? The Republicans are framing the whole debate."

The White House's strategy was not readily apparent beyond the West Wing. "We wanted to manage that first exchange," Axelrod later explained.

Obama and his staffers knew these cuts were never going to become reality; they would be blocked in the Democratic-controlled Senate. As one White House official put it, "If the cuts were heading to the president's desk, we'd be out there fighting them. . . . We had a strategic goal, and we were willing to take a short-term hit."

There was a lot of hit-taking going on. The president was being lashed by deficit hawks for not unveiling a comprehensive tackle-the-debt scheme. He was disappointing progressive policy advocates by not assailing the GOP's effort to shred the government. His approval rating among independents was down from the previous month.

Obama could not publicly explain the strategic reasoning behind these decisions. If he acknowledged openly that his budget was merely a tactical placeholder or conceded that he did not consider the CR the main event, he would undermine his own negotiating position in each of these battles—and enrage, respectively, the deficit-first crowd and his progressive allies.

Whether or not these were the right tactical moves, one thing was certain: it was not possible to have a presidency with explanatory footnotes.

OBAMA WAS JAMMED IN ANOTHER WAY. HE AND HIS ECONOMIC advisers feared that a government shutdown could trigger a severe economic downturn by shaking the markets and undermining consumer and business confidence. They would have to reach a deal, just as they had during the tax debate in December 2010.

At one meeting with the president, Lew and other members of the economic team gave him the big picture: the House Republicans were driven by their $100 billion promise and to achieve that they would have to decimate the programs Obama cared about the most, including Head Start and education programs. But Lew, who knew his way through the weeds of the federal budget, pointed out that if the cuts could be held to less than half that amount, he could structure the reductions so the impacts would not be severe and the president's favorites could be protected.

You won't have to choose between funding for the National Institutes of Health and Head Start, Lew told him.

Go ahead, the president said.

ON MARCH 1, THE HOUSE REPUBLICANS AND SENATE DEMOcrats worked out a temporary compromise on an interim spending bill that would finance the government for two weeks beyond March 4 and would trim $4 billion from this year's spending (which included rescinding unspent money for water projects and economic initiatives). Two days later, as Biden, Daley, and Lew were motorcading to the Hill to hold the first session of budget talks with the Republicans, Gene Sperling and Dan Pfeiffer came to the White House briefing room to announce that Obama was willing to cut $6.5 billion from his budget to meet Boehner halfway.

The math was a bit complicated. As Sperling explained it, the CR that funded the government through March 4—which Obama signed—was $40 billion below the president's budget request for 2011. The recent, short-term extension had included $4 billion in cuts. And with the new $6.5 billion in reductions—which Sperling would not detail—the president had reached $50 billion in cuts, which was half of the $100 billion the Republicans had promised.

But House Republicans were claiming that their budget measure sliced $61 billion from current spending levels—not from Obama's larger 2011 budget request, which was never enacted. From their perspective, Obama was counting a $40 billion cut that didn't actually exist and accepting only $10 billion in additional cuts.

So was it $100 billion versus $50 billion, or $61 billion versus $10 billion?

This focus on the numbers (open to debate) showed how difficult it would be for the Obama administration to avoid the who's-cutting-more story line and push a growth message.

At this briefing, John Harwood, a reporter for CNBC and the *New York Times,* asked Sperling if the president was ratifying the GOP's central operating premise that cuts were good economic policy.

"You're saying substantially that [the Republican] premise is correct," Harwood inquired, "that by definition cuts are not anti-stimulative, and, in fact, the [confidence-boosting] benefits could outweigh the . . . loss of particular jobs?"

Sperling sidestepped, noting that spending cuts might indeed yield "confidence benefits"—meaning more business investment and consumer demand—but if "done too deeply" they could have a "negative impact" on the economy. He was not willing to specify the "exact level" of spending cuts that might be expected to harm the economy.

ON CAPITOL HILL, AFTER THE MEETING BROKE UP BETWEEN White House officials and congressional leaders, Boehner took Reid aside.

"We don't need these guys," the speaker told the leader, referring to the White House gang. Boehner thought he and Reid could bang out an accord on their own. They were both legislators who operated in the old-fashioned style: work out a deal that makes everyone unhappy, then shake hands.

Reid nodded and remembered the advice Representative Steny Hoyer had recently given him about Boehner: He's a nice guy, you'll enjoy working with him, but he can't close the deal.

As negotiations gingerly moved forward, Boehner and the Republicans remained fixated on downsizing federal agencies and programs *now*, with the Tea Party House members yearning for even deeper reductions than the whopping cuts already approved by the House.

In the Senate, moderate Republicans were complaining that their House comrades had gone too far. And moderate Senate Democrats, including several worried about reelection the following year, were complaining that a Democratic countermeasure to the draconian House-passed CR did not cut enough. In the upper chamber, both that Senate Democratic alternative and a GOP bill reflecting the House Republicans' cuts failed. There was a stalemate.

Jack Lew and Rob Nabors, who now headed the White House legislative office, were trekking regularly to Capitol Hill to meet with Boehner's chief of staff, Barry Jackson, and Reid's chief of staff, David Krone. But this was all hush-hush. One day, Capitol Hill reporters spotted Lew and Nabors meeting with Boehner staffers in one of Reid's offices, but none of the participants would confirm that any substantial talks were under way.

The mood was tense at a mid-March lunch meeting of Senate Democrats and David Plouffe, Jack Lew, and Gene Sperling. Several of the lawmakers were upset that the president had been doing what they perceived as his above-the-fray thing in the past few weeks, while they'd been in the trenches trying to beat back the Republican antigovernment crusade. They were also worried that Obama would undercut them and yield too much on spending cuts in these private talks with Boehner's staff.

The venting at the lunch continued: Obama wasn't shaping the debate; he wasn't pounding the Republicans. When Senator Al Franken didn't receive a sufficient reply to a question about the status of the budget talks, he shouted at Sperling, "Give me a goddamn answer. Just give me a fucking answer."

The administration officials were reluctant to disclose any of the concessions Obama or Boehner were discussing for fear they would quickly leak to the media—and these proposals would immediately become targets. Tea Partiers, in particular, were poised to go ballistic over any sign that Boehner was being reasonable.

"Progress was being made in the talks," a senior Democratic Senate aide later said. "But it was all so fragile. Every time Boehner's staff talked to Reid's staff, the eighty-seven House Republican freshmen and the Tea Partiers would get upset and ask, 'Why are you even negotiating?'"

ON MARCH 11, OBAMA HELD A PRESS CONFERENCE TO ADDRESS spiking oil prices and to discuss the earthquake and tsunami that had rocked and flooded Japan. He encouraged Republicans and Democrats in Congress to reach a compromise, but he declared there were GOP proposals he would not accept, such as cutting Pell grants and bouncing kids out of Head Start.

Nor would he stand for the dozens of legislative riders Republicans had attached to the bill dictating various policies. These measures would block the Obama administration from implementing its health care overhaul, from regulating certain pollutants and greenhouse gas emissions, from setting up the Consumer Financial Protection Bureau, from creating a consumer products complaints database, and from providing money to Planned Parenthood for (nonabortion) gynecological services it made available to low-income women in health clinics across the country.

The president was finally engaging. The Republicans, he contended, were exploiting the budget battle to advance their favorite conservative causes.

"My general view is," he said, "let's not try to sneak political agendas into a budget debate."

These weren't quite fighting words, but Obama's line in the sand was real, if not boldly etched.

With the negotiations proceeding in a low-profile manner, the House passed another extension setting April 8 as the now-we-mean-it deadline and slicing another $6 billion from 2011 spending. The vote count was a warning to both Boehner and the White House. The speaker lost fifty-four Republicans—a combination of Tea Party freshmen and veteran right-wingers. Without them, he needed Democratic votes to pass this crucial bill, and eighty-five House Democrats obliged—only to avoid a shutdown. Though Boehner and other GOP leaders had urged all the House Republicans to hold together, close to a quarter of his caucus wanted to threaten a shutdown to force Senate Democrats and the White House to accept deeper cuts.

This episode rankled House Democrats who felt they had saved Boehner's hide but were still being cut out of the main action—the Capitol Hill–White House negotiations. When Bill Daley met with the House Democratic caucus that day, he was forced to listen to complaints about Obama being disengaged and waiting to leap in at the end to resolve congressional bickering. Daley noted that Obama didn't want to come across as "too shrill" and maintained that a replay of Clinton's victorious 1995 assault on the Gingrich Republicans was not possible, given the present circumstances.

IN THE SECRETIVE NEGOTIATIONS, A PATTERN SET IN. THE Republicans would complain that the Democrats were not cutting enough (the White House was now offering about $20 billion in reductions below current spending levels); the Democrats would complain that the Republicans kept shifting their position because Boehner was under pressure from Tea Party Republicans who were against any compromise and wouldn't mind if the government came to a grinding halt.

At the same time, Senate Democrats were worried that a freelancing Daley might hand Boehner more cuts than they fancied and undermine their negotiating posture.

Obama, in accordance with the White House plan, was not saying much in public. Robert Reich, in one of his many television appearances, pleaded for Obama to spring into action: "The president is the only one who has the bully pulpit. If he doesn't counter these Republican mistruths, then he is going to be in trouble next year."

In the final week of March, Boehner's negotiators suggested that the speaker might settle for about $35 billion in cuts—instead of $61 billion—if Obama would yield on several of the riders. The White House signaled it could go as high as $33 billion. (The discretionary budget would end up near $1.05 trillion, so this difference was budgetary crumbs.)

On March 30, Vice President Biden spoke to Boehner and tried to lock in the $33 billion level. Boehner refused to agree until the riders were resolved. That day, at a closed-door Republican caucus meeting, Boehner, speaking of the Democrats, boasted that he would "kick their ass."

Reid, for his part, was ticked off at the White House for accepting such a high number. He wasn't happy that Daley had pushed the talks along by apparently offering more cuts to Boehner. "Daley was going rogue," a senior Democratic Senate aide recalled.

The next day, Reid publicly declared that the Democrats and Republicans had agreed to $33 billion. Boehner, though, refused to acknowledge any agreement. "Nothing will be agreed to until everything is agreed to," the speaker said, and insisted he was still fighting for the $61 billion measure.

Reid and his negotiators knew Boehner had consented to $33 billion but that he couldn't dare admit this in public without being assailed by his fellow Republicans. Cantor was waiting in the wings to lead a rebellion if Boehner lost control of his caucus.

THE PRESIDENT DIDN'T HAVE THE LUXURY OF PONDERING ONE knotty dilemma at a time. The fire hose was always turned on. Crises hit (uprisings in the Muslim world, a Japanese nuclear meltdown), and issues and events were always looming. Forget multitasking; this was hypertasking. And despite the rush of current problems, the president always had to be looking ahead. So as the dustup over the current spending levels continued, Obama was plotting the next phase of the fight: a big deficit plan that would be accompanied by a call for bipartisan negotiations to resolve the nation's entire fiscal situation.

Sperling was holding meetings to craft a major deficit-reduction proposal. He organized subgroups to ponder the various pieces: Medicare and Medicaid, tax revenues, discretionary spending, mandatory spending. It was tough to do this while the CR negotiations were proceeding and a possible government shutdown neared. But there was a saying in the White House: Don't be the soccer ball. That is, kick rather than be kicked.

Sperling believed that any deficit-reduction framework Obama made public would have to contain enough specifics to be seen as credible—without giving away too much before the negotiations.

Eventually, Sperling and his colleagues began presenting

options to Obama, who had no problem immersing himself in the eye-glazing details. Both the Bowles-Simpson commission and Ryan's budget had claimed about $4 trillion in savings over ten years. Sperling and his economic team did not believe that amount of cutting could be achieved without awful consequences. They wanted it stretched out over a longer period.

In one of these meetings, Sperling proposed the same target of $4 trillion but over thirteen or fourteen years. He believed his formula would show that the administration was serious, but it would not risk placing a drag on the economy or require harsh cuts in Medicare and Medicaid.

"If you make it twelve years, that would be better," the president said.

He phrased it as a suggestion, but that was his way of giving an instruction. Twelve years it would be.

Obama began thinking about a speech.

THE LONGER THE BEHIND-THE-SCENES BUDGET TALKS DRAGGED on, the more antsy Obama became. He had thought the negotiations would take a few weeks, not more than a month. The Republicans seemed to be dragging it out. He wanted to address their assault on government spending; he was eager to contrast his vision of government with theirs. He also wanted to get on to the big discussion—about long-term deficits, the national debt, taxes, and investment.

We have to make this about values, he told his aides.

WITH A GOVERNMENT SHUTDOWN ALMOST A WEEK AWAY, THE negotiators on all sides picked up the pace. Lew was pushing for the deal to include reductions in mandatory programs, such as farm subsidies. Changes in mandatory programs—CHIMPS, in budget-wonk lingo—was a longtime congressional practice that often produced savings by reducing funding that might not actually occur. Resorting to CHIMPS allowed Lew to protect other parts of the budget and prevent a deep cut in the baseline for discretionary spending. He knew additional cuts were coming, and he didn't want to let the Republicans lower that baseline too far in this first round.

On Monday, April 4—four days before the shutdown deadline—Boehner met with the Republican caucus to prepare his troops for the final days of battle. He told them he was upset about what had happened two weeks earlier, when fifty-four of them voted against him on the last short-term extension. He had felt abandoned, he said.

The chastising didn't go over well with the Tea Partiers, some of whom complained that Boehner was being too emotional. But the speaker was acknowledging that he wasn't able to discipline Republicans who were not willing to follow him into a deal.

At this raucous meeting, the speaker warned his comrades that they could end up the losers if the government shut down. "The Democrats think they benefit from a government shutdown. I agree," he said. (White House aides were not sure of this.)

Veteran GOP lawmakers who had been around during the 1995 budget clash backed Boehner on this point. But Boehner's radicals were not convinced a shutdown would hurt their party. Some argued it could help. They were willing—even eager—to take that risk.

Tea Partiers and other conservatives pressed Boehner to play hard—to stick with the $61 billion. They didn't care about ancient history; they weren't scared of Obama, or the media . . . or anything. They had not come to Washington to split the difference. They believed Obama and the Senate Democrats could be vilified for resisting what they saw as the overwhelming message of the past election: cut it all.

Boehner didn't have full command of his troops. This meant Obama was negotiating with someone who might not be able to cover whatever check he wrote.

EARLIER THAT DAY, OBAMA'S REELECTION CAMPAIGN MADE IT official by zapping out an e-mail from "Barack" to millions of his supporters declaring that the president was filing papers to launch his 2012 reelection bid.

"We've always known that lasting change wouldn't come quickly or easily," the e-mail noted. "It never does."

It was an odd day to make such an announcement. Not only was the president negotiating a compromise that would yield budget cuts guaranteed to enrage Democrats and progressives, but that day, Attorney General Eric Holder announced that Khalid Sheikh Mohammed, the 9/11 mastermind, and four alleged coconspirators would be tried before a military commission, not in a civilian court.

News reports said the military commission trial would likely occur at the Guantánamo detention facility—a reminder Obama had not made good on his campaign vow to close the prison that had come to symbolize the excesses of George W. Bush's national security policies.

Liberals saw the president's inability to shutter the prison as emblematic of a series of civil liberties letdowns. And with this declaration, the Obama administration was essentially raising a white flag on Gitmo.

Neither Obama nor Holder was happy about this. At a press conference, Holder said he still believed KSM should face judgment in a civilian court. (His plan to try KSM and the others in a federal court in New York City had been scuttled after local politicians raised a fuss.) He angrily decried members of Congress for passing legislation to block the trials of detained suspected terrorists in civilian courts.

"As the president has said, those unwise and unwarranted restrictions undermine our counterterrorism efforts and could harm our national security," Holder declared.

The president and the attorney general had lost the battle of Gitmo. For that, Obama was accused by some of having broken his promise, which was not a fair criticism. He had not abandoned his pledge or flip-flopped. The Guantánamo episode was an unfortunate case study in politics trumping national security—in which Obama never figured out how to counter and overcome demagoguery from both parties.

In his first days as president, Obama had signed an executive order to shut down Gitmo within a year. It was hardly a controversial action. John McCain had called for the facility to be closed, and President Bush had publicly stated his desire to see it be shuttered.

"When we were drafting the executive order, we thought, maybe naively, that the Republicans were on board to close it down," a senior administration official at the time recalled.

Then Gitmo became a political grenade. That spring—after the White House decided to start releasing seventeen Chinese Uighurs held at Gitmo, who had been swept up in Afghanistan but deemed not enemy combatants—Representative Frank Wolf, a Virginia Republican, protested, exclaiming this would "directly threaten the security of the American people." (The first Uighurs were to be released, under strict conditions, in northern Virginia, into a Uighur community.)

Wolf's complaint ignited another conservative assault on the president, with critics claiming Obama was intent on importing dangerous terrorists into the United States. Other Republicans enthusiastically jumped into this scare campaign, extending their grandstanding beyond the Uighurs' case to cover Obama's plans to relocate Gitmo detainees to one or more high-security facilities in the United States. (They forgot the inconvenient point that high-security prisons in the United States already held dozens of terrorists—none of whom had ever escaped to harm Granny down the street.)

Democrats, too, went along with this fact-free ride. In May 2009, Senate Democrats joined with Republicans to block $80 million to fund the closing of Guantánamo and to prohibit the administration from transferring Gitmo detainees to the United States except for trial. Reid told reporters he didn't want terrorism suspects in the United States. And the White House hardly fought back.

"We were getting calls from the Hill: 'Do you care about this?' " a White House official at the time recalled. "The signal received, if not sent, was that closing Gitmo was not a top priority."

Greg Craig, the White House counsel who was leading the Gitmo effort, pressed on, looking for a home in the United States for the Gitmo detainees. The administration identified the maximum security Thomson Correctional Center, located in a rural area of northwestern Illinois, as the best spot. But to proceed, it needed funds. The Pentagon informed the White House it would not reprogram the $150 million or so to carry out the transfers without a congressional vote.

Winning such a vote—as recent events had demonstrated—would be tough, and at the time the legislative shop of the White House was obsessed with the president's health care initiative.

"Axe kept getting polls showing that either the American people did not really care about this or were against it," a White House official recalled. "We were losing the politics. And the politics had changed dramatically from before the election to after the executive order."

The Gitmo promise might have been close to dead by the summer of 2009. But Pete Rouse spent months working on a plan under which the federal government would buy the Thomson center and move the Gitmo detainees—who were declining in number—to this site. (Ten Uighurs had been transferred to Palau and Bermuda.) Prisoners considered too difficult to prosecute—in some cases because they had previously been subjected to torture or harsh interrogations—or too dangerous to release would be held without trial indefinitely at Thomson, a policy that upset civil libertarians but which Obama had reluctantly defended.

"In order to close down Gitmo, Obama was okay with a small population of detainees being indefinitely held, as long as there was a review of the evidence and the law in these cases," a senior Obama aide recalled.

Toward the end of 2009, White House officials were publicly acknowledging that they would miss the January 22 deadline for shutting down Gitmo, but they insisted progress was being made and Gitmo would close in the not-too-distant future. In mid-December, Obama signed a presidential memo ordering Gates and Holder to acquire the Thomson prison as the replacement for Guantánamo—though it was clear that fearmongering legislators in Congress would keep blocking this project. (Civil liberties groups also opposed the Thomson proposal, citing the continuation of indefinite detentions.)

The one-year anniversary of Obama's promise came and went without official notice—though a Justice Department task force at that time concluded that about 50 of the 196 Gitmo detainees ought to be held indefinitely without trial under the laws of war (but with a revamped review process). "No one knew what to do with these people," a senior Justice Department official said.

"These cases had been fucked up by the Bush administration. There was no obvious solution."

White House officials realized that conditions at Guantánamo had dramatically improved and were within international norms. But they were aware that the symbolic significance of shuttering Gitmo remained. Congressional Democrats, though, were telling White House aides they had zero interest in an election-year fight over Gitmo.

The House Armed Services Committee, a panel controlled by Democrats, voted unanimously in May 2010 to prohibit the administration from establishing any facility within the United States as a replacement for Gitmo. Weeks later, the *New York Times* reported the death knell for Obama's Guantánamo promise: "Stymied by political opposition and focused on competing priorities, the Obama administration has sidelined efforts to close the Guantánamo prison, making it unlikely that President Obama will fulfill his promise to close it before his term ends in 2013."

At the end of 2010, during the lame-duck rush of legislating, Congress went further and passed a measure barring the Pentagon from transferring Gitmo detainees to the United States for any reason—including prosecution. That seemed to be checkmate on Gitmo. The legislation doomed Holder's uphill effort to bring KSM to justice in New York City.

"There was only so much Obama could do given the convergence of forces," recalled a former White House official who worked on the issue. "We had spent hours and hours on this. We had explored options and options. Obama had a strong civil libertarian streak. But he was operating in a world of limited political options."

Holder had little choice. If KSM were to be tried at all, it would have to be at Gitmo before a military commission. Obama and his aides had not been able to plot a path through the shoals of political demagoguery and institutional obstructionism.

In April 2011, announcing his surrender on the KSM trial, Holder blasted members of Congress for ginning up "needless controversy" and "scoring political points." He added, "It is still our intention to close Guantánamo." But what had once been a top-drawer Obama promise now seemed a hollow refrain. The simplistic view was that Obama had not kept his word. The actual story was he had failed to overcome fierce bipartisan opposition.

HOURS AFTER HOLDER'S STATEMENT, AS PART OF HIS REELECTION announcement, Obama held an evening conference call with hundreds of campaign volunteers and organizers. The president emphasized his accomplishments that played well with the progressive base: repealing Don't Ask/Don't Tell and winding down the war in Iraq. He also addressed liberals who might be disappointed about compromises on health care, Wall Street reform, and budget bills, and his failure to pass climate change legislation, enact immigration reform, and shut down Guantánamo.

"In a democracy," the president observed, "change takes time. If there are some things we have not gotten done yet, it's because we have not yet been able to mobilize the American people around these issues."

He told his loyalists they "should never be defensive" about this administration's record.

That day Holder's throw-in-the-towel announcement was a vivid reminder of the gap between politics and governance, of the difficulties of implementing the best of intentions, and of the obvious fact that not every fight was winnable.

On Guantánamo, Washington politics and popular indifference had trampled Obama. With the unresolved budget battle before him, Obama was hoping to show he could at least fight to a decent draw.

Chapter Eight

LEADING FROM BEHIND
THE SCENES

ON MARCH 15, 2011, PRESIDENT OBAMA AND HIS TOP AIDES SET-
tled into their seats in the Sit Room for a National Security Council meeting. The forces of Libyan leader Muammar al-Qaddafi
were on a roll. The ragtag rebels seeking to topple the dictator
had wrested control of Libyan cities on the Mediterranean. But
now Qaddafi troops were gaining momentum, and casualties
were mounting.

The White House was receiving chilling accounts of the violence transpiring in areas seized by Qaddafi loyalists, and government forces were poised to overrun Ajdabiya, a town that
supplied fuel and water to Benghazi, the rebel stronghold in the
east. If Qaddafi's soldiers succeeded there, Benghazi would likely
fall next. And if Qaddafi took Benghazi, the uprising could be
crushed. Worse, a massacre could follow. A bloodbath loomed.

The president asked how soon Qaddafi's troops could be in
Benghazi. The answer: forty-eight to seventy-two hours.

Prior to the meeting, various options had been prepared for

the president, including US forces supporting a proposed no-fly zone over Libya that was being pushed by European allies. Obama turned to Mike Mullen, chairman of the Joint Chiefs: "Would a no-fly zone do anything to stop this scenario?"

"No, sir," Mullen replied. "It wouldn't stop the advance of Qaddafi's forces."

The government troops were using tanks and ground-launched weapons against the rebels, not air power. Preventing Qaddafi's warplanes from attacking would do little to protect the rebels and civilians in Benghazi.

"So why are we discussing this?" Obama asked.

The president set aside the papers that had been drafted for the meeting. None of this, he said, will stop possibly thousands of people from being butchered on our watch.

"Give me real options," he said.

Obama knew there was a divide within his administration. Traditionalists did not favor military action, believing there was no paramount US interest at stake. Others, particularly several senior NSC staffers, saw Libya as a case demanding military action designed not to advance a realpolitik interest but to prevent a slaughter.

Robert Gates noted that he was not keen to have another war in a Muslim nation. He was already managing two. He wondered how the United States could limit its engagement in Libya. He and others at the Pentagon worried they would be forced to shift resources from Afghanistan, which would make it harder to stick to a planned drawdown of US forces there.

Mullen also was not in favor of intervention—nor was Biden, Donilon, and McDonough. They all supported some sort of response but were skeptical of direct US military involvement. Some aides asked, Did Obama want to be sucked into another military action when his number one job was to do something about the ailing economy?

Hillary Clinton and Susan Rice, the US ambassador to the United Nations, backed aggressive action. Clinton's earlier public comments had suggested she did not support military intervention, but recently she had been pushing internally for a more assertive stance.

Clinton, who was calling into the meeting from Egypt, believed that Libya's instability could interfere with the democratic transitions in neighboring Egypt and Tunisia. Permitting Qaddafi to smash the uprising in his country, she argued, would send the wrong message to Iran. She also said she'd recently received assurances that Arab governments, particularly the United Arab Emirates, were willing to participate in military action against Qaddafi.

This time the Americans would not be barreling into another Muslim country on their own. During the Egyptian crisis, the Saudis had constantly complained to administration officials that Washington was unduly throwing Mubarak out. Now the message from the region was different: feel free to get rid of Qaddafi for us.

Rice, speaking via a video connection from New York, urged direct attacks on Qaddafi's tanks and artillery. She had served on President Clinton's NSC in 1994 when the United States did little to stop the genocide in Rwanda. Months after the mass killings there, she visited the country and walked through a churchyard where a massacre had occurred. Decomposing bodies were still strewn about. Stepping around those corpses would always be a searing reminder for her.

In her first major speech as UN ambassador, Rice declared the administration's support for Responsibility to Protect—R2P, as foreign policy mavens called it. This principle holds that governments have a fundamental obligation to protect their populations from genocide, war crimes, ethnic cleansing, and the like—and that other states must take collective action if a nation cannot meet this fundamental responsibility. Qaddafi had killed thousands of his own citizens over his forty-two-year rule. It was clear what would happen to the people of rebellious Benghazi.

The massacres of Somalia, Bosnia, Rwanda, and Darfur were on the ambassador's mind. After all that, she had resolved that in any similar situation she would speak her mind so that no matter what happened she could sleep well afterward. "I would come down on the side of dramatic action," Rice once said, "going down in flames if that was required."

As far as Rice sussed it out, the president's national security team was split into a do-nothing camp and a do-something camp. She believed that the options presented to the president had been designed to prevent rapid action. But Obama had impressed her by shoving that all aside and moving to the fundamental matter: Would the United States use its mighty but overstretched military to prevent a bloody nightmare?

Rice proposed seeking a UN Security Council resolution that would go beyond endorsing a no-fly zone and authorize the United States and other countries to mount direct military assaults against Qaddafi's war machine. Winning approval of the measure would be tough—the Russians and Chinese might block it—but she was eager to try.

Obama then polled the backbenchers—the aides who didn't sit at the table. Not surprisingly, Samantha Power backed military action. In 2005, when Obama was a freshman in the Senate, he had asked Power, then a Harvard professor, to meet with him. He had read *A Problem from Hell,* her Pulitzer Prize–winning history of the US government's generally tepid responses to genocides in the twentieth century. During a conversation that lasted hours, the two bonded, and Power soon joined Obama's Senate staff.

In the White House, Power was a leader among aides who favored muscular but multilateral idealism. In early December 2009, when Obama was about to fly to Oslo to accept the Nobel

Peace Prize, he summoned Power at the last minute to journey with him on Air Force One so she could work on his unfinished acceptance speech.

Her mark on that address was clear, at least in one passage: "I believe that force can be justified on humanitarian grounds, as it was in the Balkans, or in other places that have been scarred by war. Inaction tears at our conscience and can lead to more costly intervention later."

That had been theory—this was the real thing. Power was nervous about what might happen if military action led to the fall of Qaddafi's regime. There was a risk that fundamentalists or al-Qaeda would be empowered. Yet the United States and its allies could not let thousands of people die, if such a tragedy could be prevented.

Obama also asked Tony Blinken, Biden's national security adviser, for his opinion. Blinken disagreed with his boss—for whom he'd worked nearly a decade—and said he favored military action. It was a brave move. Ben Rhodes also sided with the advocates for intervention.

The president wasn't surveying these aides because he was having trouble making up his mind. He was leaning toward intervention, and he was looking for more voices that reflected his own view.

Later on, published accounts would contend that Obama had been nudged toward intervention by the women on his national security team: Clinton, Rice, and Power. But there was more of a generational divide than a gender split among his advisers. Rice, Power, Blinken, and Rhodes were a younger breed of foreign policy wonks, each now driven more by policy beliefs than the institutional concerns that Gates and Mullen had to consider.

Obama told his aides he had not come to the White House to be a leader who turned the other way. He noted that the United States was not in a position to intervene whenever a massacre

loomed somewhere in the world. But in this instance it could—
and with backing from the region.

Obama departed the meeting with an order: "I want to see
all the options."

He meant military options.

THE TEAM REASSEMBLED LATER THAT EVENING AND PRESENTED
three options to the president. The first, crafted by McDonough,
was essentially what the president had called for: seek a UN Security Council resolution calling for all necessary measures to
protect the citizens of Libya. The United States would let the UN
and its allies know this meant striking at Qaddafi's forces on the
ground and that the United States was willing to participate in
the assault. The second option was to join the French and British
in a no-fly zone. The third was to stay out.

Obama went around the room again. Gates, Mullen, and
Biden questioned whether this was vital to US national security. Gates pointed to the history of these sorts of endeavors: the
United States joins, believing its role will be limited, but ends up
bearing much of the burden.

James Steinberg, the deputy secretary of state (who was calling in for Clinton), noted that the secretary's recent discussions
in Europe, the Middle East, and North Africa had convinced her
that the Europeans and the Arab countries would support direct
military action. Rice echoed her earlier comments. The advocates
for a forceful response maintained that this was not just a matter
of preventing a humanitarian crisis. If Qaddafi were permitted
to smash the uprising, the Arab Spring would lose momentum.

The president summed up his view: standing by is not who
we are. The United States could not say no when presented with
this unique alignment of circumstances: a preventable bloodbath, an invitation to intervene from Libyans, and support from

other governments in the region. He definitely wasn't going to back an ineffectual no-fly zone, which he considered little more than a political ploy: flying planes over Libya, while Qaddafi was butchering people below. Despite the apprehensions of his two top military advisers, his vice president, and his national security advisers, Obama was willing, on the strengths of his own beliefs, to commit the United States to a new sort of war.

If events went badly, the warnings of Obama's top national security advisers would, no doubt, become public, and Obama would end up on the wrong side of a we-told-you-so tale. He could pay a tremendous political price. During the Egyptian crisis, Obama had disregarded the calls for caution from his much-experienced senior national security advisers. He was now doing the same. This was a daring decision.

Obama asked Mullen to draw up plans for NATO military action in Libya. He instructed Rice to seek a broad resolution from the UN Security Council that would allow the United States to join with other countries to protect Libyan rebels and citizens by attacking government forces. Obama was insistent that this operation could only proceed with UN approval and Arab participation—and no American boots were to touch Libyan soil. There was little, if any, talk of an endgame or exit strategy. The aim was to stop a disaster—and then work out what ought to occur.

The president was moving deliberately—pulling along much of his national security team—and, as his aides would put it, Obama was once again being "forward-leaning." Having once hailed hardheaded realism and questioned the wisdom of deploying American troops to avert humanitarian disasters, the president was now being propelled by a humanitarian impulse. Strategic considerations were not absent from his decision. Clinton was correct that events in Libya could impact actions in Egypt and elsewhere. But this was about doing the right thing.

A month earlier, on February 11—the day that Mubarak gave up power—critics of Qaddafi announced that peaceful demonstrations would be held the following week. But quickly the uprising turned into a civil war.

In the initial days, the White House responded cautiously— even as events moved quickly, the death toll increased, Libyan officials began to defect, and calls came from home and abroad for Obama to endorse regime change.

The White House released a statement from Obama condemning "the use of violence by governments against peaceful protesters" in Libya—and also Bahrain and Yemen, where other antigovernment actions were under way. But Obama and his aides did not want to turn the Libyan uprising into a Washington-Tripoli face-off or—with several dozen Americans stationed at the US embassy in Tripoli, and hundreds more Americans living in Libya—trigger a hostage crisis.

Behind the scenes, administration officials hurriedly pulled together options: freezing assets, imposing international sanctions, ostracizing Libya in international organizations. When Obama finally spoke on the crisis—saying "this violence must stop" and reporting that his administration was preparing "the full range of options"—it was a statement with no details and little oomph.

Nearly two weeks passed—a veritable lifetime in this day of 24/7 constant Internet and cable coverage—without a forceful public response from the White House. Critics chastised Obama for not vigorously calling for an end to Qaddafi's reign.

But once the Americans were out of Libya, the administration flipped a switch. The president signed an executive order imposing unilateral sanctions on the Qaddafi government, and he declared that Qaddafi had "lost the legitimacy to rule" and should leave immediately. The assets freeze Obama ordered was no small matter; it resulted in Treasury Department officials seizing $30 billion in Libyan funds—the largest seizure of funds in US history.

For weeks, the main public debate was whether the United States should join the no-fly zone Britain and France were pushing. Administration critics on the right and the left accused Obama, who showed no eagerness to sign up for this mission, of dithering. Yet inside the White House, aides favoring military intervention and those who did not actually agreed that a no-fly zone did not make much sense. Gates and company believed it would come with high costs. Rice, Power, and others viewed it as inadequate for stopping Qaddafi's forces.

But the White House was taking other actions. It intensified humanitarian relief operations delivering material to rebel-held areas. Intelligence officials and diplomats were trying to reach Libyan military and government officials to persuade them to turn on the Libyan dictator. American officials were making contact with the Libyan opposition. Still, at a White House press briefing on March 7, Jake Tapper repeatedly asked Carney "how many people would have to die" before Obama would launch military action.

As the Libyan opposition pleaded for the international community to move faster—and Arab states were voicing support for a no-fly zone—administration officials were telling reporters that Obama was content for the time being to let other nations take the lead in divining a solution. "This is the Obama conception of the US role in the world—to work through multilateral organizations and bilateral relationships to make sure that the steps we are taking are amplified," Rhodes told the *Washington Post*. "Maybe this is a different conception of US leadership. But we believe leadership should galvanize an international response, not rely on a unilateral US response." In a March 10 conference call with reporters, Tom Donilon noted, "We're pursuing a range of military options"—without convincing reporters that the White House was yet serious about such action.

At a March 11 press conference, the president was asked if he was prepared to "use any means necessary" to clear Qaddafi from power. Obama gave no indication he would soon be planning an attack on the Libyan dictator. "We are slowly tightening the noose," he said.

Slowly—that was the key word in Obama's response. As Qaddafi's troops were brutally retaking rebel-held towns and heading toward Benghazi, the president, perhaps caught between those aides calling for immediate action and those who questioned whether military intervention in Libya was in sync with US national security interests, appeared to be taking his time.

BUT AT THE MARCH 15 SIT ROOM MEETING, OBAMA LEAPED over the no-fly zone debate and was now pushing for war.

The next day, Rice worked the UN and made it known she had been in a meeting with the president the night before and he had decided to pursue a "robust" response. She pressed for a Security Council resolution that would endorse military strikes against Qaddafi's military to prevent it from murdering civilians. The United States, Rice told other diplomats, would commit to full participation in the front end of the action, then it would be up to NATO and the Europeans to manage the operation. And she was leaning on Arab nations to commit to the mission.

European diplomats at the UN, though, thought Rice was bluffing—and trying to torpedo the British and French call for a no-fly zone. They believed her proposal for military intervention above and beyond a no-fly zone was a scheme designed to get

Obama off the hook. Washington, they assumed, was promoting this overly ambitious resolution only because Obama and his aides believed that it would fail. When this effort crashed and burned, Obama would then be able to hold up his hands and say, "Well, we tried"—and would be free of any obligation.

The Europeans were misreading the president. But perhaps they could be forgiven for such cynicism, given that the White House was not trumpeting the fact that Rice was pushing for such a forceful resolution. At a press conference in Cairo, Clinton lowballed what Rice was up to, merely noting that Rice was working for a "resolution that would include a range of actions the international community could take." Yet that evening, Mullen and Donilon presented Obama with plans for air strikes on Libya.

Obama made it clear to his aides: he was not prepared to use force unilaterally. He needed the UN stamp of approval. One White House staffer asked if aides should prepare a backup plan to assemble a coalition of the willing, including France, England, and a couple of Arab states, in the event Rice failed at the UN. The answer was a firm no. The president was counting on Rice.

At the UN, Rice lobbied Portugal and Brazil—two of the nations with seats on the Security Council. And she was focused on Russia. Moscow and Beijing tended to follow each other's lead, and Russia was the nation to watch on Libya. China would likely vote as Russia did.

Obama helped by calling South African President Jacob Zuma; his nation held one of the fifteen seats on the council and had long been cozy with Qaddafi.

"This is a personal priority," Obama told Zuma.

Clinton, then in Tunis, granted Ryan Lizza of *The New Yorker* a breakfast interview by the hotel pool. She indicated that the president had decided to intervene. She noted that it had not been easy to reach this point: "I get up every morning and I look around the world. People are being killed in Côte d'Ivoire, they're being killed in the Eastern Congo, they're being oppressed and abused all over the world by dictators and really unsavory characters. So we could be intervening all over the place."

She contended that obtaining regional support was instrumental: "For those who want to see the United States always acting unilaterally, it's not satisfying. But, for the world we're trying to build, where we have a lot of responsible actors who are willing to step up and lead, it is exactly what we should be doing." This would be intervention Obama-style, not Bush-style—the result of careful deliberation and multilateral diplomacy, supported (if not called for) by the people overseas who would be most affected.

That evening, Rice won the resolution she and Obama desired. It called for a no-fly zone and "all necessary measures" to protect civilians. That meant that European and Arab nations willing to intervene in Libya could take almost any sort of military action they thought best against Qaddafi's military apparatus. Rice had performed a masterful feat of diplomacy. She won ten out of the fifteen votes on the Security Council (including South Africa's) and managed to prevent the other five—China, Russia, Germany, India, and Brazil—from opposing the resolution; these nations abstained.

Not only was Rice setting up a never-before-seen US-European-Arab military alliance to intervene within an Arab country on humanitarian grounds; but by forcing a resolution embodying the Responsibility to Protect principle out of the Security Council, she was establishing a powerful precedent.

Her success surprised officials throughout the government. "Nobody thought they could get 'all necessary measures,'" recalled a senior Senate staffer following the issue. The top lawyers in the Justice Department, who would have to contend with various legal issues generated by any military action in Libya, were caught off guard. "We were shocked when the Russians didn't veto it," a senior Justice Department official later said. "Many thought this wouldn't happen, and we wouldn't have to worry about the legal issues that might come up. But now, it was, 'Oh shit, we're in Libya.'"

AFTER THE EVENING VOTE AT THE UN, OBAMA AND HIS NA-tional security team met to discuss military options. He approved permitting US warplanes and navy vessels to join European and Arab nations in the initial strikes against Qaddafi. But US in-volvement in the campaign had to be limited. "Days, not weeks," he told his advisers. Though it was late, he called both British Prime Minister David Cameron and French President Nicolas Sarkozy to make sure they understood that the United States would not, per Gates's fear, end up holding the bag.

That day, Qaddafi had vowed that his forces would soon be in Benghazi. "There won't be any mercy," the Libyan dictator declared, adding, "We will wipe out this black page of our his-tory." He called on his supporters "to go out and cleanse the city of Benghazi."

THE NEXT DAY, THE PRESIDENT MET WITH A BIPARTISAN GROUP of congressional leaders in the Situation Room; several lawmak-ers phoned in. The participants included John Boehner, Mitch McConnell, Harry Reid, Nancy Pelosi, Richard Lugar, John Kerry, Jon Kyl, Eric Cantor, and other Democrats and Repub-licans who oversaw military and intelligence matters. Obama walked them through his thinking.

The point, he said, was to stop a massacre—with the full support of the international community and the Arab League—for the obvious benefits and to protect space for the Arab Spring to continue. He insisted this military action would be limited, with no US troops entering Libya. We're not targeting Qaddafi, Obama said, but the ultimate goal was to force him from power.

The lawmakers expressed few objections, though Lugar said that he did not support the strike. "I didn't hear anyone complain that he should have given more notice," Senator Carl Levin, the Democratic chair of the Armed Services Committee, later said.

Shortly after that, Obama went on television and said that without military intervention, "many thousands could die" in Benghazi and that the "entire region could be destabilized, endangering many of our allies and partners."

No ground troops would be deployed, he said, adding, "We are not going to use force to go beyond a well-defined goal—specifically, the protection of civilians in Libya."

Change in the region, he asserted, "will not and cannot be imposed by the United States or any foreign power; ultimately, it will be driven by the people of the Arab World. It is their right and their responsibility to determine their own destiny." He presented Qaddafi with an ultimatum: unless he implemented a cease-fire, stopped the advance on Benghazi, and pulled his troops out of Ajdabiya and other towns, "consequences" were coming.

That night, Obama met with Gates, Mullen, Clinton, Donilon, Daley, and other administration officials to review the military's final plans. Afterward, the president and his family headed to Air Force One for an overnight flight to Brazil for trade talks. Clinton flew to Paris to tend to the diplomatic tasks necessary for this new military action. Obama's aides knew questions would be raised about Obama leaving town as a military action was commencing. But the president had often told his aides that he had to project the ability to do two or more things at once.

The next day was proof of that theory. In the morning, Obama had an hour-long meeting with Brazilian President Dilma Vana

Rousseff. Afterward, he met with Donilon, Daley, and Rhodes and received a secure call from Mullen and Gates.

Has Qaddafi shown any signs of stopping? Obama asked. Was Qaddafi responding to yesterday's ultimatum?

No, they reported.

Obama ordered the launch of the attack.

Soon afterward, US and British ships and submarines fired more than 110 Tomahawk cruise missiles at twenty Libyan air defense positions. French warplanes were over Benghazi. NATO and coalition forces attacked military assets the Libyan Army could use against Benghazi. Qaddafi's advancing troops were stopped.

THE WAR WAS ON—BUT THE ADMINISTRATION REFUSED TO CALL it that. On *Meet the Press,* Mullen would only refer to it as a "limited operation" in support of "humanitarian efforts protecting the civilians in Libya."

But he acknowledged that Qaddafi's forces would be hit to "ensure they are unable to continue to attack the innocent civilians."

Naturally, the military operation would spark political warring in Washington. Immediately, liberal House Democrats questioned the constitutionality of the missile strikes. Republican Senator John Cornyn tweeted that the president was treating Congress as a "potted plant." John Boehner issued a have-it-both-ways statement, declaring that the United States "has a moral obligation to stand with those who seek freedom from oppression and self-government," but noting that Obama had to "better explain" the mission in Libya.

Obama sent a letter to Congress, notifying both houses that the strikes will be "limited in their nature, duration, and scope"—and thus were within his authority to commence on his

The day after the 2010 midterm elections, which President Barack Obama called a "shellacking," he shares with his aides a long list of ambitious legislative goals for the short lame-duck session. His aides wonder, as one says, "What does he see that we don't?"

With less than three weeks left in the congressional session, Vice President Joe Biden joins Obama in an Oval Office meeting to discuss strategy for negotiating a tax-cut deal with Republicans. Biden, unlike Obama, relishes haggling with their former colleagues in the Senate.

Obama works on a statement with economic adviser Gene Sperling, senior aide Mona Sutphen, and message strategist David Axelrod to announce a tax-cut compromise with the GOP. Democrats in Congress would accuse Obama of caving to Republican demands to extend tax breaks for the rich, but with the accord Obama craftily wins a second stimulus.

Obama signs legislation on December 22, 2010, repealing the Don't Ask/Don't Tell ban on gays and lesbians openly serving in the military. He plotted a careful and deliberate political strategy to achieve this win—and considers this day one of the best of his presidency.

After the Senate ratifies the New START treaty, Obama, communications director Dan Pfeiffer, Axelrod, and press secretary Robert Gibbs head from the West Wing to the Eisenhower Executive Office Building for a press conference, where Obama hails "the most productive post-election period we've had in decades."

Obama shares a champagne toast with aides who worked to ratify the New START treaty. Minutes later, boarding Air Force One for a holiday vacation, Obama would be smiling and singing "Mele Kalikimaka," the popular Hawaiian Christmas song.

As he prepares for the third year of his presidency, Obama meets with (*from left to right*) senior adviser Pete Rouse (*behind Obama*), senior adviser Valerie Jarrett, counsel to the president Bob Bauer, new White House chief of staff Bill Daley, senior adviser David Plouffe, and deputy chief of staff for operations Jim Messina. The president would start this year with a revamped—and more strategic-minded—staff.

Obama takes part in a conference call in the Situation Room to discuss the shooting of Representative Gabrielle Giffords and others in Tucson, Arizona. Joining him are (*from left to right*) national security adviser Tom Donilon, Daley, Messina, Pfeiffer, and legislative affairs director Phil Schiliro. Four days later, at a memorial service, Obama would call for civility and humility in the national political discourse.

Obama greets House Speaker John Boehner before delivering his State of the Union speech on January 25, 2011, in which he would call for greater government investments in education, innovation, and infrastructure. Obama would spend much of the coming year negotiating with Boehner, who would have a tough time handling the Tea Party extremists within the House Republican caucus.

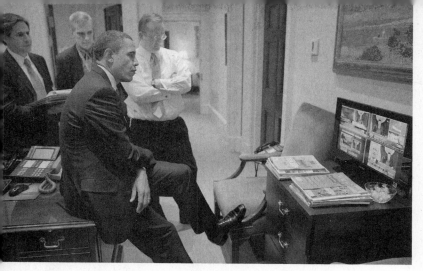

Days into the uprising in Egypt, Obama, joined by Tony Blinken, the vice president's national security adviser; Denis McDonough, the deputy national security adviser; and Gibbs, watches President Hosni Mubarak deliver a post-midnight speech. Shortly afterward, Obama would call Mubarak and urge him to implement extensive democratic reforms quickly before issuing a public statement declaring the United States would "stand up" for the rights of the Egyptian people.

Having lost patience with Mubarak, Obama—with Gibbs pacing about, Daley on the couch, and Donilon huddling with McDonough and Ben Rhodes, the deputy national security adviser for strategic communications—calls the Egyptian ruler on February 1, 2011, and encourages him to resign. After the call, Obama would publicly urge an orderly transition in Egypt "now."

At a meeting in the Situation Room on March 15, 2011, Obama, despite the hesitation of top national security advisers, decides to seek aggressive multilateral military action to stop Libyan dictator Muammar al-Qaddafi's troops from overrunning resistance forces and possibly slaughtering civilians in the rebel stronghold of Benghazi.

While in Brazil for trade talks, Obama launches the attack on Qadaffi's forces. Here he takes a break from the scheduled meetings to draft a statement on Libya with Daley, Donilon, and Rhodes.

As the minutes tick down to a possible government shutdown on April 8, 2011—and Boehner's team is still not making the deal final—Obama roams the West Wing, ducking in and out of meetings, carrying a football. Here he's on the phone with Boehner.

After a last-minute budget deal averts a shutdown, Obama boasts that the compromise includes the "biggest annual spending cuts in history." Though White House negotiators slyly lessened the impact of the cuts, Democrats worry the deal will bolster the Republican argument that spending cuts are the key to prosperity.

Days after the near shutdown of the government, Obama reviews a fiscal policy speech with White House legislative affairs director Rob Nabors, Treasury Secretary Tim Geithner, National Economic Council director Gene Sperling, Office of Management and Budget director Jack Lew, and chief speechwriter Jon Favreau. The address would slam Republicans for embracing draconian spending cuts, seeking to end the Medicare guarantee, pushing tax breaks for the rich, and advancing a "deeply pessimistic" vision of the nation's future.

This photo of Obama, Biden, and aides tracking the progress of the Osama bin Laden raid became an iconic image, but the White House would not disclose what the president and his team were watching when the photograph was taken. *From left to right, seated:* Brigadier General Marshall "Brad" Webb, McDonough, Secretary of State Hillary Clinton, Secretary of Defense Robert Gates. *From left to right, standing:* Admiral Mike Mullen, Donilon, Daley, Blinken, counterterrorism aide Audrey Tomason, chief counterterrorism adviser John Brennan, director of national intelligence James Clapper.

The White House released this composite of images from the meetings that took place in the Situation Room the day of the bin Laden mission so "people might have a better sense of what it's like in presidential meetings of historic significance."

As Biden-led talks to raise the debt ceiling falter—due to Republican opposition to increasing revenues or taxes—the president plays golf with Boehner. They are joined by Biden and Ohio GOP Governor John Kasich. The golf match would lead to secret talks between Obama and Boehner.

Over the July 4 weekend, Obama holds the second of his secret meetings with Boehner in pursuit of a grand bargain: a politically tough deal that would raise revenues and reduce entitlement spending. In public, though, Boehner is bashing the president for not leading on budgetary issues.

When Tea Party–driven Republicans refused to raise the debt ceiling without spending cuts, the White House's economic team spent months negotiating with the GOP to avoid a default. Here Obama discusses those talks with (*from left to right*) Daley, Nabors, Biden chief of staff Bruce Reed, Sperling, senior economic aide Jason Furman, Lew, Plouffe, and Geithner.

After Boehner privately agrees to pursue a grand bargain, the president convenes a series of White House meetings with congressional leaders to assemble the $4 trillion deficit-reduction package. Pictured here with Obama (*from left to right*): House majority leader Eric Cantor, House minority leader Nancy Pelosi, Boehner, Senate majority leader Harry Reid, and Senate minority leader Mitch McConnell.

Days after walking away from the grand bargain, Boehner revives talks with the White House, and he and Cantor meet with Obama and Biden in the Oval Office. Afterward, their aides begin hammering out the details—until Boehner renounces the deal again.

Just two days before a possible US government default, Obama strikes a deal with congressional Republicans to raise the debt ceiling. It will avoid a near-term replay of the crisis but includes none of the "shared sacrifice" Obama had wanted. Here the president works on the statement announcing the agreement in the Outer Oval Office.

As soon as the debt ceiling ends, Obama asks Sperling to craft a major jobs package that will be big and bold. Here he is with senior aides in the Roosevelt Room, days before unveiling the package. *From left to right, seated:* Daley, domestic policy adviser Melody Barnes, deputy senior adviser Stephanie Cutter, Jarrett, Pfeiffer, deputy chief of staff Nancy-Ann DeParle, Lew, senior aide David Lane, and Sperling. *From left to right, standing:* economic aide Brian Deese and Plouffe.

Moments before addressing a joint session of Congress—during which Obama would unveil his American Jobs Act bill and adopt a more confrontational approach toward the Republicans—he chats with Cantor at the entrance to the House chamber.

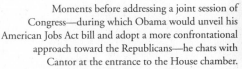

Obama fiercely campaigns for the jobs package—particularly in swing states—to put pressure on Republicans and persuade voters that his top priority is boosting employment and economic growth. Here he attends a rally in Cincinnati, Ohio, beneath an aging bridge connecting the home states of Boehner and McConnell.

At a time when Obama stopped seeking compromises with the Republicans and is confronting them on jobs, taxes, spending, and the role of the government, he tours the new Martin Luther King, Jr. Memorial in Washington, DC.

After weeks of campaigning against Republicans—who blocked his jobs bill—Obama improves marginally in opinion polls, giving White House aides hope that he is on the right track. On the upswing, Obama and First Lady Michelle Obama attend the Carrier Classic basketball game held on the flight deck of the USS *Carl Vinson*.

Toward the end of the year, Obama delivers a speech in Osawatomie, Kansas, where he declares "this raging debate" isn't "about class warfare. This is about the nation's welfare."

At a holiday reception, Obama interrupts a photo line to confer with Nabors about the payroll tax cut extension. In the last showdown of the year, House Republicans threaten not to accept a bipartisan compromise passed in the Senate. But Obama and other Republicans would assail their obstructionism, causing Boehner to yield and hand Obama his final win of the year.

One of Obama's goals for 2011 was to set up the 2012 campaign as a choice between progressive values and Tea Party extremism. He largely succeeded in doing so. But with the economic recovery remaining anemic and the public mood sour, he ended 2011 still facing great obstacles to reelection.

own say-so. But during a press briefing, Donilon was asked to square the administration's actions in Libya with Obama's declaration in 2007 that a president could only authorize this sort of attack with Congress's prior approval. Obama noted that "consultation with Congress is important" and "the administration welcomes the support of Congress in whatever form that they want to express that support." It was not an answer.

Over the next few days, the assaults on Qaddafi's troops continued, and Obama and his aides worked to ease the transition to a NATO command.

As Obama continued his trip through Latin America, with stops in Chile and El Salvador, he was frequently briefed on the military operations and was often on the phone with allies to sort out the details of the command structure of the coalition. He called the emir of Qatar to make sure this Arab nation (along with the United Arab Emirates) was participating in the Libyan attacks.

Secretary of State Clinton was vigorously keeping the coalition from unraveling, focusing on tensions between Italy and France that had developed on the first day of the attack when French warplanes attacked a column of tanks hours before the official start of the assault. She also had to pressure the UAE, Qatar, and Jordan to keep their military commitments to the alliance, after these Arab states got cold feet following the start of the bombing campaign.

Though polls showed public approval of the Libya mission, Boehner tried to paint Obama as a rogue and attempted to boost the impression that the mission had "a lack of clarity." He sent the president a sharply worded letter: "It is regrettable that no opportunity was afforded to consult with Congressional leaders" before Obama ordered the Libya attack. Yet Boehner had been part of the pre-assault briefing of congressional leaders and had raised no fuss. And the point of the mission had been clearly defined: stop Qaddafi from committing a massacre.

On March 23, the White House press office, sensing that this did not seem to be sinking in, offered Denis McDonough as a guest to all the cable networks. On *Hardball,* host Chris Matthews dubbed Libya "a war without explanation," and he asked, "The Obama doctrine—can you define it? We can't."

A tired-looking McDonough—the bags beneath his eyes were quite dark—explained that Obama, facing a potential slaughter, took practical and deliberate steps to avert it: "I can't give you an ideology behind it. What I can tell you is that when we're presented with this kind of [threatening] language that Qaddafi presented, seeing the kind of history that he brings to this kind of threat, and knowing that this is a moment of great opportunity in the region, [the president] thought it was a good opportunity for us to take."

Matthews responded, "You basically are arguing common sense."

THE OUTCOME IN LIBYA WAS STILL UNCLEAR. THE UNITED States and its allies were targeting military assets in a manner that could lead to the overall defeat of the Libyan dictator. But Obama's initial achievement was the prevention of a bloodbath in Benghazi and perhaps elsewhere—and getting political credit for something that doesn't happen is always tough.

An exasperated Clinton noted at the time, "I know the nightly news cannot cover a humanitarian crisis that thankfully did not happen, but it is important to remember that many, many Libyans are safer today because the international community took action."

Still, the president, now back in Washington, had to keep on the offensive in the messaging battle. In his weekly address, he said the Libyan campaign served the nation's strategic interest by preserving stability in the region. Gates, the skeptic, remarked in public that though events in Libya did not create an immediate threat to the United States, a massacre there could have negatively affected the course of the Arab Spring. He and Clinton, in media interviews, acknowledged the military campaign in Libya could drag on for weeks or months.

A few days later, Obama, speaking at the National Defense University, again tried to shape the narrative. "If we waited one more day, Benghazi, a city nearly the size of Charlotte, could suffer a massacre that would have reverberated across the region and stained the conscience of the world," he said. "It was not in our national interest to let that happen. I refused to let that happen."

He answered the critics who said he had moved too slowly: "When people were being brutalized in Bosnia in the 1990s, it took the international community more than a year to intervene with air power to protect civilians. It took us 31 days." NATO, he said, would take command of the operation within days. (Obama sidestepped the matter of the United States' dominance within NATO.)

The speech was a ringing endorsement of pragmatic humanitarian interventionism. Obama contended that critics who had accused him of inconsistency—for engaging in intervention in this instance, but not others when human rights were threatened—had missed the point: "It's true that America cannot use our military wherever repression occurs. . . . We had a unique ability to stop that violence [in Libya]: an international mandate for action, a broad coalition prepared to join us, the support of Arab countries, and a plea for help from the Libyan people themselves."

But he insisted that the mission was not merely about

do-goodism. There was an "important strategic interest" propelling the military campaign: "A massacre would have driven thousands of additional refugees across Libya's borders, putting enormous strains on the peaceful—yet fragile—transitions in Egypt and Tunisia. The democratic impulses that are dawning across the region would be eclipsed by the darkest form of dictatorship, as repressive leaders concluded that violence is the best strategy to cling to power."

Obama was talking in grand terms—but recognizing limitations. He also used this moment in the speech as a riposte to neoconservatives. The Iraq war, he noted, had not been worth the massive costs. This time, the United States had assembled a truly broad coalition and won UN approval for the military intervention.

If a pundit were looking for an Obama doctrine, perhaps it could be boiled down to this: when a nightmare looms, you do what you can, when you can, considering cost and benefits and taking into account the potential for coalition making and regional attitudes. That might not be poetic, but it was typical Obama, blending the noble and the practical.

IN THE COMING WEEKS, NATO RATCHETED UP THE ATTACKS, with US warplanes and Predator drones occasionally participating in assaults, in order to apply more pressure on Qaddafi. CIA operatives and British special forces were on the ground in Libya, assisting the rebels and gathering intelligence on the best targets for NATO warplanes to hit. (Obama had signed a secret order authorizing covert US support for the opposition forces.)

A *New Yorker* magazine article posted on May 2 sparked a fuss by quoting an unnamed Obama adviser who described the Libya operation as an example of the president "leading from behind." Hawkish critics seized on this phrase as a fitting and derogatory summation of Obama's weakness. His primary role in pushing the Europeans, the Arabs, and the UN beyond the likely-to-be-irrelevant no-fly zone was still not widely recognized or credited.

In mid-May, Obama hit a deadline—and a constitutional controversy. Under the War Powers Resolution of 1973, passed in response to the Vietnam War, a president must terminate a military action after sixty days if Congress has not authorized it. And Congress had not okayed the Libyan operation. In the thirty-eight years since the measure was passed, it had often been overlooked—by both lawmakers and presidents. Legislators didn't want to stop military actions; nor did they want to own them. Presidents did not fancy acknowledging that Congress could restrict their commander-in-chief powers.

Obama, who had once taught constitutional law, was in a predicament. As a senator, he had clearly stated that a president could not initiate military action without Congress's assent, unless it was an act of self-defense. Now, in the middle of this military action, he did not want to cede power to Congress, especially the Republican-led House.

Administration lawyers studied the issue and came back with a split decision. The Justice Department and the Pentagon concluded that the Libyan operation was governed by the War Powers Resolution, and Obama required authorization from Congress to continue the mission; otherwise he had to terminate the operation or scale back US participation.

But the State Department and the White House determined that the military action in Libya did not rise to the level of hostilities covered by the War Powers Resolution. That is, it really wasn't a war—at least, not for the United States after NATO took command and the United States moved to mainly a support role. There were no troops on the ground, and few, if any, in harm's way. That meant Obama could proceed without seeking a green light from the House and the Senate.

Both findings were presented to the president, and not surprisingly, he preferred the one that permitted the war—or the operation—to continue and that allowed him to duck taking a stand on the constitutionality of the War Powers Resolution.

Lawmakers of both parties were enraged.

"The president is not a king, and he shouldn't act like a king," bellowed Representative Dan Burton, a conservative Republican from Indiana.

Representative Brad Sherman, a liberal Democrat from California, insisted, "It's time to stop shredding the US Constitution in a presumed effort to bring democracy and constitutional rule of law to Libya."

Some supporters of the Libyan action—including John McCain, John Kerry, and Joseph Lieberman—called for Congress to pass a resolution endorsing the war, which would end this debate. But many Capitol Hill lawmakers were more comfortable blasting Obama than saying one way or another if they supported or opposed the military action against Qaddafi.

This was a typical congressional maneuver: don't take a stand on an operation that might turn out badly, but preserve the right to criticize. Early in the operation, Boehner had let Obama know that Republicans would neither raise a fuss about the mission nor afford it legitimacy by voting in favor of it. The speaker, according to a senior administration official, had essentially said, "Don't drop this on our lap."

It was an uncomfortable spot for Obama. "So what does the president do?" a Justice Department official later asked. "Tell NATO, France, and England, 'Sorry, something has come up'? By all accounts no one thought this would last until May 20. In March, the lawyers were telling the policy people, 'You have only sixty days,' and the policy people were saying, 'Okay.' They weren't worried. It was just a lawyer's issue."

Obama had a choice: pull out of the operation, which would end the NATO operations and create a geopolitical mess, or stand with the lawyers who said the military action was not "hostilities," knowing he would take a political hit for that. Obama, who had come into office hailing the rule of law, was now engaged in what seemed a slippery constitutional endeavor—though White House lawyers insisted their interpretation of the much-disputed War Powers Resolution was as solid as the more restrictive opinion of their Justice and Pentagon colleagues.

Obama, however, would not adopt the view embraced by the Bush-Cheney administration that the commander in chief has total power to wage war. Obama's position was nuanced: he didn't challenge Congress's right to oversee the war-making authority of the chief executive; he just claimed this particular military action was not subject to that authority. Some of his lawyers agreed; some did not. It didn't matter what the Obama of 2007 might have said.

As the military campaign proceeded, on June 3, 2011, the House passed a resolution rebuking Obama, with most Republicans and forty-five Democrats supporting it. Another measure calling for a withdrawal of US forces from the Libyan operation failed 148–265, but with sixty-one Democrats backing it.

The War Powers Resolution argument was not ending. Senator Dick Durbin, an Illinois Democrat and faithful ally of the president, said Obama's position "doesn't pass a straight-face test in my view that we're not in the midst of hostilities." In the meantime, McCain blasted House Republicans for trying to end the Libya operation.

The politics were a whirlpool that culminated in another odd set of votes in the House. On June 24, the chamber voted against legislation authorizing the Libya mission on a 295–123 vote, with 70 out of 190 Democrats siding with the Republicans. A second bill, strongly endorsed by Boehner, to prohibit any direct US involvement in combat operations also failed, on a 238–180 vote. Eighty-nine Republicans deserted Boehner on that vote.

The House wouldn't authorize the operation, but it wouldn't limit it either. In the Senate, McCain, Lieberman, and Kerry couldn't get a vote on their measure to approve the mission. It all made for a confusing picture, which helped Obama avoid an all-out political donnybrook.

At a press conference, the president poked at Congress for fixating on this matter: "We should be sending a unified message to [Qaddafi] that he should step down and give his people a fair chance to live their lives without fear. And this suddenly becomes the cause célèbre for some folks in Congress? Come on."

With much else under way in Washington—most notably, the start of a furious debate over raising the debt ceiling—and the partisan divide on Libya hazy, the question of the legality of the Libyan operation never produced a full-fledged commotion. The controversy faded; the war continued. Then near the end of August—after five months of NATO pounding Qaddafi's forces—the rebels took Tripoli, and the Libyan regime toppled. Qaddafi could not be found, and his loyalists still held other parts of the nation. But this seemed the end for the dictator who had seized power in 1969.

"The future of Libya is in the hands of the people," Obama said.

His approach—moving cautiously and deliberately (too slowly for some) toward a truly multilateral intervention predicated on humanitarian concerns—appeared to have worked. But Obama would be tested in other countries undergoing popular uprisings, notably Bahrain, Yemen, and Syria. The ruling governments in Bahrain and Yemen had been allies of Washington; Bahrain hosted the US Navy's Fifth Fleet, and Yemen had cooperated in a key counterterrorism program.

Obama had not fully supported the protesters in these countries. As for Syria, administration officials were no fans of President Bashar Assad, but they did fear his government could be succeeded by a more hostile regime, perhaps Sunni fundamentalists. In a May 19 speech reflecting on the Arab Spring, Obama had made a big statement: "Our message is simple: if you take the risks that reform entails, you will have the full support of the United States." That could be a hard pledge to redeem consistently and completely across the region.

In Libya, Obama had started a war of protection—which had turned into a war of liberation—while enlisting other countries to share the burden. He had not been leading from behind. Perhaps he had been leading from the side—or from behind the scenes. But he had authored this operation, and he had chased Qaddafi out of power, without resorting to full warfare.

In October 2011, Qaddafi would be captured in his hometown of Sirte. He would be beaten and abused before being killed. His decomposing body would be put on display in a meat locker and then buried in a secret desert location.

The day Qaddafi was killed, Obama went to a podium in the Rose Garden to tell the nation, "We're under no illusions—Libya will travel a long and winding road to full democracy. There will be difficult days ahead." The Arab world was remaking itself; Obama was encouraging this transformation. As one White House aide said, "There was no way of predicting with certainty what would be the end of any of this."

Chapter Nine

The End of the Beginning

In the afternoon of April 5, 2011, President Obama made an unscheduled appearance in the White House briefing room. He was exasperated.

That morning he had met with John Boehner and Harry Reid in the Oval Office. Obama and Reid both thought that the three of them had agreed on $33 billion in spending cuts for the continuing resolution needed to fund the government through September.

But Boehner issued a new demand: $40 billion. The speaker apparently had not been able to sell the $33 to the conservatives and Tea Party freshmen of his caucus. Obama argued that he had already gone more than halfway to the $61 billion in the Republican House budget bill. Now Boehner was backtracking, when they were three days away from a government shutdown?

Standing before the journalists, Obama made a case for sealing a deal quickly: the last thing the economy needed was "a disruption" caused by a government shutdown. He said he would not agree to another short-term CR extension (unless it was only to provide time to finalize an agreement).

He denounced the riders and decried the process: "What we're spending weeks and weeks and weeks arguing about is actually only 12 percent of the budget, and is not going to significantly dent the deficit or the debt."

Americans, he added, expect us to "act like grownups."

Obama was sharp but not personal. He didn't assail Republicans by name. Boehner was not as kind. That day, he blasted Obama for "not leading on this year's budget"—even while the two of them were engaged in intensive talks to craft a deal.

The next night the speaker was back in the Oval Office—with Reid, Biden, and the president—again wrestling over the final amount of spending cuts and the legislative riders. The atmosphere was tense.

"Nothing will be agreed to, until everything is agreed to," Boehner reiterated. He offered $39 billion.

Obama felt he was being jerked around. The number had been $33 billion, then $40 billion, and now $39 billion. The Republicans were holding the government hostage *and* exploiting the situation, tacking onto this must-pass bill provisions that otherwise could not get through Congress. Obama had wars to oversee, uprisings in the Arab world to handle. He had to work out a deal. He and his aides truly believed that the economy could be roiled by a shutdown.

AFTER THE MEETING, WHITE HOUSE OFFICIALS, REID AIDES, and Boehner staffers met through the night, until 3:00 A.M., with the Democrats countering Boehner's last offer with a package of about $34.5 billion in cuts. Each side, though, had different ways to tally reductions—Jack Lew was looking for as many CHIMPS as he could get—and there was disagreement over funding for the Pentagon.

As for the riders, Obama was willing to consent to the less

consequential add-ons. But he was holding firm against the provisions aimed at health care reform, environmental protections, and Planned Parenthood.

Later that morning—this was the day before D-day—Boehner defied the president by bringing to a vote a stop-gap bill that would fund the military for the rest of the fiscal year and keep the rest of the government open for one more week. Obama had vowed to veto the measure. This was a sideshow, perhaps designed by Boehner to bolster his position among Republicans suspicious of his private dealings with the president. He was juggling two imperatives: forging a deal and retaining the backing of his caucus so he could remain speaker.

That afternoon, Obama, Boehner, and Reid gathered once more in the Oval Office. Much of the talk was about the riders, not the big numbers. The three found themselves in the weeds, haggling over mountaintop mining rules. Was this any way to run the government of the most important economy of the world? The meeting ended with no agreement.

Through the day, Democrats were complaining that the policy riders were the main obstacle at this stage, especially the Planned Parenthood measure. "We are very close on the cuts and how we make them," Reid told reporters. "The only things—I repeat, the only things—holding up an agreement are women's health and clean air."

BOEHNER'S PROBLEM, IT TURNED OUT, WAS NOT ONLY THE TEA Partiers; it was the veteran social conservatives within the Republican conference. A Tea Party tussle had morphed into yet another skirmish in the decades-old culture wars.

The main policy rider threatening to sink everything was the anti–Planned Parenthood measure that would kill $327 million in Title X family planning funds for contraception services, cervical cancer screenings, annual gynecological exams, sexually transmitted infections testing, and other health services. This program, signed into law by President Richard Nixon in 1970, assisted millions of low-income women who obtained these services at forty-five hundred clinics across the nation, some run by Planned Parenthood. With this rider, antiabortion foes were lashing out at the nation's leading abortion provider. Obama had tried to duck a battle over Republican cuts only to become mired in the never-ending abortion fracas.

That day, Representative Mike Pence, a leading social conservative in sync with the Tea Party, took to the House floor to chant, "Shut it down!"

IN THE EVENING, OBAMA, BOEHNER, BIDEN, AND REID WERE back in the Oval Office—their third huddle in a day.

Boehner said he needed a big top-line number to sell an agreement to his fellow Republicans. There appeared to be agreement now on $38.5 billion in cuts. The negotiations had been more difficult because the White House had insisted on having a say over the specific cuts. Obama had pushed for that—so Lew would be able to protect the programs the president cared about most.

Obama and Reid told Boehner they would not yield to the GOP's attempt to defund Planned Parenthood. Obama, though, was willing to make a painful concession.

"John, I will give you DC abortion," he said, referring to a rider that prevented the District of Columbia government from using nonfederal funds to pay for abortion care—a restriction Republicans had previously imposed on the capital city.

Boehner pushed the president on the Planned Parenthood rider.

"Nope, zero, this is it," Obama said.

But as this session was drawing to a close, Boehner tried again on the Planned Parenthood rider.

Biden lost his temper and stood up.

If Boehner didn't relent, the vice president said angrily, "We're going to have to take it to the American people."

The ability of the federal government to function during wartime appeared to hinge on subsidized pap smears.

Before Boehner left, he said, "Mr. President, if you don't mind, I want to go and pray about this one."

THE GOVERNMENT WAS NOW ALMOST TWENTY-FOUR HOURS from shutting down. The White House canceled a trip Obama was scheduled to take to Indianapolis the next day.

As staffers met on Capitol Hill to go over details and once more crunched numbers into the wee hours of the night, Lew became suspicious that the Republicans were not abiding by the terms agreed upon by the president and the speaker.

The GOPers were insisting on a number higher than $38.5 billion, and they were still talking about the Planned Parenthood rider. Around 3:00 A.M., Lew told Barry Jackson, Boehner's chief of staff, they seemed to be going around in circles, with the Republicans trying to reopen issues already settled in the Oval Office. Lew went home, believing the government would close at midnight.

ON FRIDAY MORNING, *FIRST READ*, A NEWSLETTER PUT OUT BY NBC News, noted, "The question Dem and GOP lawmakers need to ponder: Are they really going to shut down the federal government over $300 million to Planned Parenthood? That's what the debate has really come down to."

The mainstream media narrative was not that Republican

extremists were on the warpath. It was whether or not Washington could work. A *Washington Post* article seemed to blame the president, asking if Obama was out of touch: "Is he diffident in the face of serious challenges?" The piece charged Obama with "procedural caution" and "an innate distaste for ideological confrontation," noting that Obama had raised "doubts about the strength of his leadership." Republican antiabortion fanatics were forcing a shutdown showdown, yet the problem was supposedly Obama's leadership style.

For good or bad, Obama was demonstrating strategic discipline, hewing to the blueprint Plouffe had designed *and* staying true to his top policy priorities: preventing any dislocation to the economy that could be caused by a government shutdown and defending programs that were part of his growth agenda. He was also taking a stand on the Planned Parenthood and environmental riders.

Obama realized he could be bashing the Republicans for their extremist cuts and policies and for their threats to crash the government. But he had calculated that doing so would lessen the chances of reaching a shutdown-preventing accord. As the drama built on this deadline day, he ordered his aides to refrain from assailing the Republicans—to preserve political space for the negotiators to nail down a deal.

By late morning, Lew and Nabors were stationed on Capitol Hill, catching and throwing offers and counteroffers with Republican staffers. The Republicans abandoned the EPA riders. The White House believed that $38.5 billion was the final number. But the Republicans were still pressing for more. Obama had to call Boehner to ensure the negotiators stuck with the $38.5 billion.

"I am the president of the United States and you are the speaker of the House," he told Boehner. "We are the two most consequential leaders in the US government. We had a discussion last night and the staff negotiations don't reflect that. If we have an agreement, it has to stick. We will have to do business for the next year and a half from our different perspectives."

The number stuck, and by midafternoon, the work was done, a provisional deal had been reached—or so it seemed to Lew. He headed back to the White House, anticipating an announcement by 5:00 P.M., with seven hours to spare.

Then nothing happened. Lew waited hours for the final word from Boehner.

As evening came—and the minutes toward midnight ticked away—Lew and Biden were in Daley's office at the White House, with Lew constantly on the phone with Nabors, who was waiting for an answer from Boehner. Republican staffers were telling Nabors there were still a few unresolved issues.

On another line, Lew was talking to federal officials about how to proceed with the government shutdown.

At one point, Nabors was asked to meet with Republican aides on the House Appropriations Committee. After listening to them for a few minutes, he got the impression they wanted to reopen issues already decided.

"I don't have time for this," Nabors told Barry Jackson. "Either make a deal or not."

WITH THE MIDNIGHT DEADLINE APPROACHING, BOEHNER started a meeting with his Republican caucus at 9:45 P.M. in the Capitol basement. He didn't have a deal, but he started describing what a deal *might* look like.

Forty-five minutes later, Reid's chief of staff David Krone, Nabors, and Jackson—meeting several floors above the Republicans—shook hands. They had an agreement. It was 10:30 P.M.

The final deal was $38.5 billion below current spending levels. The Planned Parenthood rider was dead; the District of

Columbia abortion provision remained. Almost half the cuts were in CHIMPS, with several billions in savings coming from programs that would in all likelihood not spend the funds being cut. Lew had succeeded in protecting the baseline for discretionary spending. He also had secured Obama's top priorities in education and other investments. The White House had also wrung $3 billion out of the Pentagon budget.

In the basement, Boehner, still with his caucus, announced, "There's a deal." The room cheered.

In a televised statement, Obama noted, "Some of the cuts we agreed to will be painful. Programs people rely on will be cut back. Needed infrastructure projects will be delayed." But he boasted that he had achieved "the biggest annual spending cuts in history," as if that was what he had set out to do.

The president had been forced to cut important programs beyond which he believed prudent, yet he was now embracing what he had been opposing. This was not winning the future. It was beating back a Republican assault by accepting a portion of it—in an attempt to protect the economy and seize higher ground for the next and more consequential round of combat.

Obama walked over to the State Dining Room, where Biden, Plouffe, Carney, Daley, Lew, Sperling, Reed, Pfeiffer, and others were waiting for him. His aides applauded him. After a few moments, Obama returned to the Blue Room to reenact the statement for photographers. He stood at the podium and said, "This is where I pretend I'm talking." Then, with a big smile on his face, he left.

Shortly after midnight, Lew issued a memo to be circulated throughout the federal government: continue normal operation.

This was a moment of relief, not celebration. Perhaps Obama had scored points with independent voters and advanced his adult-in-the-room strategy. But the previous December's tax-cut compromise had given Obama an economic policy of consequence:

a second stimulus. This budget compromise ate up much of the president's time for the relatively small matter of financing the government for another six months. And it had reinforced the Republicans' ultimate message that cutting spending ought to be Washington's first priority.

"We became consumed with how to get through the negotiations," a senior administration official later recalled. "Our negotiators were heavily invested in getting a deal and everything became inside baseball. We were not trying to publicly win a debate. We were privately trying to win a negotiation."

The Republicans got the bumper-sticker victory, another senior administration official later observed, but the White House won on the details. As Lew saw it, the cuts probably seemed to be an eight on a scale of one to ten, but in reality they ranked maybe a two—or, at most, a three.

MEMBERS OF THE PROGRESSIVE COMMENTARIAT WERE NOT cheering. Obama and the Democrats, *Washington Post* columnist Ezra Klein wrote, were "sacrificing more than they let on. By celebrating spending cuts, they've opened the door to further austerity measures at a moment when the recovery remains fragile. Claiming political victory now opens the door to further policy defeats later."

Other liberals wondered how Obama had ended up accepting $38.5 billion in cuts, when the Republicans had initially proposed $32 billion. MoveOn blasted an e-mail to members decrying the final deal, and Justin Ruben, the group's executive director, issued a threat: "If the president and the Democrats don't stand up to Republicans, I don't see people coming out and doing the work that it would take to get them elected."

Robert Reich spoke for disheartened progressives who

believed Obama had been outfoxed by the hostage takers of the Right: "Once you start paying ransom, there's no end to it, because you embolden them." Columnist E. J. Dionne, noting that Obama's victories were largely defensive, had a sharp dig: "'Concessions we can believe in' was not the slogan Obama ran on."

Though pooh-bahs of the Right—including Rush Limbaugh—slammed the deal as a sellout for falling short of the $61 billion (or $100 billion) mark, Boehner had demonstrated a fair measure of savvy. He had been up against a Democratic president and a Democratic-controlled Senate and had to contend with the extremists within his caucus. And he managed to win about two-thirds of the spending cuts his Republican House had passed. He had forced his agenda (or the Tea Party's) upon Washington.

Boehner's challenge, though, wasn't over. The House still had to vote on the deal—and his Tea Partiers were not in a victorious mood. With the Congressional Budget Office reporting that the $38.5 billion agreement would yield only $352 million in reduced outlays in 2011, they were fuming. It looked as if Lew had snookered Boehner, which was somewhat true.

Lew was a budget policy master, and he had structured the package with minimal reductions in immediate government spending. (The CBO estimated that the package would lead to $20 billion to $25 billion in lower outlays over the next five years.) The CBO report enraged Tea Partiers on and off Capitol Hill; they were now more suspicious of Washington compromise— and of Boehner, too. The next time around, they might well be even less amenable to a deal.

—⁓—

WHEN THE BUDGET DEAL CAME UP FOR A VOTE ON APRIL 14, Boehner lost fifty-nine House Republicans who rebelled against the compromise. The bill only passed because eighty-one Democrats backed the measure.

"We were hearing from Boehner, 'We have the votes, we don't need Democrats,'" a senior House Democratic staffer recalled.

But Boehner was wrong. Nancy Pelosi opposed the bill, yet she hadn't tried to block her fellow Democrats from voting for it. The House Democrats, as angry as they were about being cut out of the deal making, were not willing to shut down the government and politically damage the Congress and the president just to stick it to the speaker. They rode in like the cavalry to save the Obama-Boehner-Reid deal. The Senate passed the bill without fuss on an 81–19 margin.

To sell the deal to conservatives, Boehner had boasted that he had ended Obama's "stimulus spending binge." But, it seemed, he was not such a master of his own caucus.

"A lot of our new members expected big change right away," a senior House Republican aide recalled. "We had to explain to them that the founding fathers had made it hard to do big changes. And up until the end, they were still complaining that we were only cutting discretionary spending and not entitlements. We'd say, 'The CR only covers discretionary spending. We've been through this a million times!'"

This did not bode well for negotiations ahead.

The first round of budget battles had consumed weeks of time that Obama had not spent advancing the grand win-the-future themes of his State of the Union address. The economic recovery was still feeble, with millions of Americans out of work. There were wars going on. Yet the government had become dysfunctional in managing itself. And Obama had not used this near breakdown in Washington to present a larger story.

"We maintained funding for the core growth agenda," a senior administration official conceded after the fact, "but did not do well articulating that publicly." The unexpected fight over the Planned Parenthood rider probably captured more of the public's attention than Obama's key budget priorities. As another senior administration official later remarked, it was not great messaging to say the GOP wanted $100 billion in cuts, and the White House preferred $25 billion.

Whether David Plouffe's overall plan was working was open to debate—even among Obama loyalists. Shortly after the deal, a former senior White House official complained that the president's top advisers were saying "this strengthens [Obama] for the bigger fight. I think the opposite is true."

This onetime Obama-ite thought the president had become overly tactical: Obama "prides himself for believing in the politics of conviction. The tragedy is he does not live up to his self-perception. The White House is trying to win back independents. They feel they overreached on health care and need to swing back. And that's a change. The president pushed health care because he believed it was the right thing to do and didn't focus on the politics. Now he has internalized a lot of the criticism."

David Axelrod dismissed talk of any Obama shift: "He has been consistent throughout his presidency; he believes that dealing with deficits in the mid- and long term is important. But he's taken care to make sure the cuts are in the back end and the out-years and won't throttle down the economy."

But had Obama allowed the Republicans to turn the national debate into nothing other than a debt seminar? "We can be faulted for this," Axelrod said later. "Nobody anticipated the degree of Republican obstreperousness and implacability. We knew there would be strident voices, but the degree to which that tail would wag the dog—we didn't assume."

It was tempting for outsiders to think Obama should have gone to war over the CR. But the president and his aides didn't see a choice. A senior administration official later lamented that being responsible for preventing adverse consequences was lonely duty in Washington.

Lew thought of the episode this way: It had been an ugly process, but it worked.

OBAMA DID NOT GET A BREATHER AFTER THE SHUTDOWN STAND-off concluded peacefully. In several weeks, the US government would hit the limit on the amount of debt it could accrue. Unless Obama pried legislation out of Congress to raise the debt ceiling, the federal government would have to stiff people and institutions to which it owed money—a long list that could include Social Security recipients, military contractors, and bondholders. It could end up defaulting.

There was no telling what that would bring, but Timothy Geithner and other administration officials didn't want to find out. They worried that a default could detonate a global financial crisis. These days, it didn't take much for financial woes to zip from one continent to the next.

In years past, Democrats and Republicans had raised the debt ceiling without much fuss. During the past decade, this had happened ten times. In 2006, Obama and all the Democratic senators voted against a hike in the debt ceiling in a symbolic gesture to protest George W. Bush's fiscal policies; but they knew the bill would pass with majority Republican support and didn't attempt to trade their votes for administration concessions.

This time, Boehner and Mitch McConnell had been telling the White House they realized how dangerous it was to play politics with a possible default. But that didn't mean they weren't going to exploit the situation—and demand severe spending cuts to reduce the deficit in return for allowing the government to cover the bills it had previously racked up. A poll at the time showed that only 16 percent of Americans favored raising the debt ceiling. And some House GOPers yearned to say no to increasing the debt limit, period, come what may.

"It was not on anyone's mind that there would be a big chunk of Republicans who wanted us to default so they could cut government spending by 40 percent," a former top White House official noted. "Reasonable Republicans were telling us that they knew we would be playing with fire if there was a fight over the debt ceiling."

But there had been warning signs. Shortly after the midterms, Austan Goolsbee was told by an acquaintance about a meeting of the incoming House Republican freshmen. They were being briefed by a prominent conservative economist and asked why they should raise the debt ceiling. The economist said it would be crazy not to—and he completely lost the room. But, the economist explained, the US government could default and not be able to pay for almost half of its services and that would put financial institutions in crisis.

So? the freshmen asked.

Goolsbee's acquaintance was passing along a warning that these new guys will be hard to work with. Goolsbee dutifully shared this with the top levels of the White House. Yet it didn't spook anyone.

"Nobody fathomed," a former White House official later said, "how out of control this could get."

WITH THE DEBT-CEILING DEADLINE APPROACHING, IT WAS TIME for Obama to head into Round Two of the budget showdown: the fight over a deficit plan. And as Plouffe had hoped for, the Republicans did go first.

On April 5, in the final and frenzied week of the 2011 spending squabble—the day Obama was haggling with Boehner over whether the budget cuts would total $40 billion or $33 billion—Representative Paul Ryan introduced the House Republicans' budget plan for 2012, which also covered spending and taxes for the next ten years. His $40 trillion blueprint was as radical as White House strategists had dreamed of.

Ryan claimed his budget would reduce the deficit by $4.4 trillion over ten years. The proposal would repeal Obama's health care overhaul and reduce tax rates for corporations and wealthy individuals. Nearly $1 trillion in discretionary domestic spending would be hacked away.

Two of the plan's most out-of-the-box provisions targeted medical entitlements. One would turn Medicaid, which covered health care for low-income Americans, from a matching program (under which the federal government pays states on the basis of the number of enrollees) into a block grant program (which would distribute a set level of funding to the states). In essence, there would be a cap on Medicaid, which would likely lead to fewer services for low-income Americans.

The other measure was truly radical: the traditional Medicare system of reimbursing doctors for health care provided to seniors would be replaced in several years by a system in which seniors would purchase a subsidized insurance plan. The Medicare guarantee would be gone, and seniors would pay on average $6,000 more a year. The savings for the feds: $389 billion.

In a way, this budget bill was a cover for the Bush tax cuts, which over the next seven years would cost about $4 trillion, close to the savings Ryan claimed. And over the coming decade, the government, under the Ryan budget, would add $5.1 trillion to the national debt. (Obama's budget called for spending that would increase the debt by $7 trillion.)

The White House had been waiting for this. Not a scuffle

over where to slice domestic spending programs for the coming six months, but a roll-up-the-sleeves confrontation over Medicare, Medicaid, and tax fairness. House Democrats eagerly pounced on Ryan's Medicare proposal, seeing this call to end Medicare's guarantee as their ticket back to the majority. The campaign ads would write themselves.

"The Ryan proposal could be the foil Obama needs," Paul Begala, the strategist/pundit, said. " . . . I hope every vulnerable Republican in Congress signs on to the Ryan plan to kill Medicare, because we will beat 'em like a bad piece of meat."

Yet the White House's immediate response to Ryan's budget was restrained. In a little over a week's time—after the CR fight was done—Obama was giving a major speech to reveal his own deficit-reduction plan. He would wait until then to address Ryan's scheme. Plouffe wanted as many Republicans as possible to jump aboard the Ryan Express before Obama moved to derail it.

FOR WEEKS, OBAMA AND HIS AIDES HAD BEEN WORKING ON THIS speech, and the president wanted a "values-based" address. It would not be a dry recitation of facts and figures defending the necessity of tightening the nation's fiscal belt. It would be the moment he had been anticipating since the end of last year, when he could directly compare his vision with what the Republicans were peddling.

In one meeting about the speech, Obama told speechwriter Jon Favreau, Gene Sperling, and other aides that a recent trip to Chile had sharpened his thinking about the address.

"I have something really important I want to say in this speech," he remarked. "I'm going to other parts of the world and they're showing me tremendous investments in infrastructure and innovations in education. They're willing to spend money on that."

Back home, he added, it was a different picture: "The Republican budget reflects a fundamental pessimism. It says that to get the deficit in line, we can't afford to be as visionary as these countries, and we can't be optimistic because they're not willing to let an extra penny come from high-income people."

The Republican vision, he added, was simply cut Medicaid and be less generous to families with disabled children—and not whip up the financial will to modernize the nation's infrastructure. Smaller countries were aiming bigger.

Obama had a lot of pent-up passion. During the budget negotiations, he had not fully spoken out. He felt the time had finally come. To draw lines, to be visceral, to call out the Republicans. He was ready to have that battle of values.

DEFICIT REDUCTION WAS NOW FULLY INTERTWINED WITH THE debt-ceiling issue. On the morning of April 13—five days after the near shutdown—Obama met with the eight top congressional leaders at the White House. With Vice President Biden, McConnell, Kyl, Reid, Durbin, Boehner, Cantor, Pelosi, and Hoyer sitting at the polished wooden table in the Cabinet Room, the president noted that he did not believe in linking the debt ceiling to spending cuts. It had never been done in the past. But tying an increase in borrowing authority to cuts had become gospel within the GOP. We understand, Obama told the lawmakers, that we have to do both at the same time. The president was yielding on practical terms without conceding the principle.

Shortly after that session, the presidential motorcade left the White House and raced the five blocks to an auditorium at George Washington University. It was time for Obama's latest Big Speech. Only a small fraction of Americans that weekday afternoon would be sitting in front of televisions airing the speech live, but Obama was hoping this address would carry far beyond the cable news reports.

At the university, Obama's audience included students, university officials, and invited notables, including Republican Representatives Paul Ryan, Jeb Hensarling, and Dave Camp, who each had served on the Bowles-Simpson deficit commission. Prior to the speech, a pro forma invitation had gone out to members of the commission, but no one expected these three to show. When Sperling saw them in the auditorium, right before Obama was to hit the stage, he was stunned. The way the room was lighted, the president wouldn't be able to see them. Their attendance, in the ways of Washington, was a conciliatory gesture, and the president was about to rip into them.

Uh-oh, Sperling thought.

Obama wouldn't have fundamentally recast his speech had he known Ryan—his chief target—and the others would be in the rows before him. But he would have added a line or two to note their presence and emphasize that this was no personal attack. In Washington, a city of big but easily bruised egos, such niceties can matter.

Not that the Republicans were always mindful of such protocols. That morning, McConnell, Boehner, and Cantor—after their meeting with the president—had held a press conference to blast Obama's speech before it was even delivered, declaring that any plan with increased taxes for the wealthy would be unacceptable.

Obama began by tracing recent history. After the nation started amassing debt "at alarming levels" in the Reagan 1980s, Presidents George H. W. Bush and Bill Clinton forged agreements with their political opponents to reduce deficits, and

"America's finances were in great shape by the year 2000. We went from deficit to surplus."

Then "we lost our way." The Bush-Cheney administration launched two wars, created an expensive prescription drug program, and enacted tax cuts without covering the costs for any of this. Obama entered office facing a $1 trillion annual deficit, and the emergency steps he had taken to deal with an economy in crisis added to the deficits. He was reminding people that the Republicans had gone wild with the federal government's credit card and now were blaming their bender on the guy stuck with the tab.

It was time, Obama insisted, to restore fiscal responsibility. But, he insisted, a deficit plan was more than just numbers; it reflected "a vision of the America we want to see five years, ten years, twenty years down the road." And then he turned to the Ryan plan—which was scheduled for a vote in the House in two days.

The GOP proposal "would lead to a fundamentally different America than the one we've known." Obama listed some of the proposed cuts: a 70 percent cut in clean energy, a 25 percent cut in education, a 30 percent cut in transportation, and cuts in Pell grants.

"These are the kinds of cuts that tell us we can't afford the America that I believe in and I think you believe in."

Obama bluntly challenged Ryan's devotion to the nation's future: "I believe [the Ryan budget] paints a vision of our future that is deeply pessimistic. It's a vision that says if our roads crumble and our bridges collapse, we can't afford to fix them. If there are bright young Americans who have the drive and the will but not the money to go to college, we can't afford to send them."

The president noted that China was opening research labs and solar facilities. South Korean schoolchildren were outpacing American students in math and science. Brazil was investing billions of dollars in new infrastructure and running half its cars on biofuels. "And yet," he said, "we are presented with a vision that says the American people, the United States of America—the greatest nation on Earth—can't afford any of this."

OBAMA BLASTED RYAN'S MEDICARE PROPOSAL, ASSERTING, "PUT simply, it ends Medicare as we know it." He maintained that up to fifty million Americans could lose their health insurance due to the Ryan budget.

"Worst of all," he continued, "this is a vision that says even though Americans can't afford to invest in education at current levels, or clean energy, even though we can't afford to maintain our commitment on Medicare and Medicaid, we can somehow afford more than $1 trillion in new tax breaks for the wealthy. Think about that."

He was charging that Ryan and his crew intended to sell out hardworking Americans to help out millionaires.

"This vision is less about reducing the deficit than it is about changing the basic social compact in America," he said. ". . . I don't think there's anything courageous about asking for sacrifice from those who can least afford it and don't have any clout on Capitol Hill."

This was no policy-wonk speech. Obama was excoriating Ryan and the other GOPers, questioning their commitment to a just and generous society. As Ryan, Hensarling, and Camp listened, they seethed.

Obama unveiled his own "more balanced approach" to reducing red ink, a "framework," not a plan, for $4 trillion in deficit reduction over twelve years. Obama would cut $770 billion from nondefense discretionary spending and save $400 billion from security programs. He proposed policy changes in Medicare and Medicaid targeting wasteful subsidies, erroneous payments, inefficient practices, and expensive prescription drugs that he said would result in $480 billion in savings—without shifting costs to seniors or low-income Americans. Obama's proposal would wring $360 billion from farm subsidies and other mandatory spending programs.

Obama would not extend the Bush tax cuts for the wealthiest, and he would pursue tax reform to close assorted loopholes. He would limit itemized deductions for the wealthiest 2 percent and pick up $320 billion over ten years. Obama proposed nothing specific regarding Social Security, noting it "is not the cause of our deficit." Altogether, his framework included $2 trillion in cuts, $1 trillion in reduced interest payments, and $1 trillion in extra revenue from tax reform.

Obama would also implement a "debt failsafe" that would trigger across-the-board spending cuts (except in Social Security, low-income programs, and Medicare), if by 2014 the ratio of debt to the gross domestic product was not declining. He called on the congressional leaders to designate members to participate in bipartisan negotiations aimed at constructing legislation embodying comprehensive deficit reduction (and lifting the debt ceiling). The talks would begin in early May and be led by Vice President Biden.

Positioning himself as the defender of Medicare while also adopting portions of the Bowles-Simpson report, Obama was blending his progressivism with deficit-oriented centrism. He said little about jobs—other than to insist that the nation's rising debt will "cost us jobs" and to reaffirm he was committed to investments that will lead to jobs in the future.

This was a forceful counterattack. Obama had waited for Republicans to act like, well, Republicans. Then he had jumped on them, claiming he was the protector of the social compact *and* a policy visionary who could leverage government investments to steer the economy toward long-term prosperity.

After weeks of the grinding CR ground war, it was an energizing moment for Obama's economic team and a reaffirmation of their core progressive values.

Was it really possible to be a liberal deficit hawk? Obama knew he couldn't beat the Republicans on the question of who wanted to cut the deepest. Voters looking for slashers would be drawn to the GOP. But if the president could change the channel from a contest over cuts to a clash over values, he might have a chance—in the upcoming negotiations and in the 2012 election.

AFTER THE PRESIDENT WAS DONE, SPERLING LOOKED FOR RYAN to tell him this wasn't personal, but he and the other Republicans left quickly.

Ryan, Camp, and Hensarling had interpreted the invitation as a peace offering, but they left feeling the address had been a personal attack on them.

A furious Ryan denounced the speech as "extremely political, very partisan." Camp, chairman of the House Ways and Means Committee, told the *Washington Post,* "What came to my mind was: Why did he invite us?" And the newspaper reported, "The situation was all the more perplexing because Obama has to work with these guys."

Boehner immediately confirmed that it would be difficult for Obama to forge a compromise, for he reaffirmed that the Republicans would not yield an inch on boosting tax revenues.

The deficit speech seemed a turning point. Obama was playing offense. With a full-throated defense of government, he was appealing to his ideological base. With his concession to the deficit worrywarts, he was courting back independent voters. A top Democratic strategist noted, "The speech against the Ryan budget was what we wanted in 2010."

Referring to the address, a senior Treasury official gushed at the time: "The best of Obama and as robust a defense of the role of government in today's world. He did the Clintonian thing—managed to make both [Paul] Krugman and the centrist types happy—more or less."

Obama also satisfied the deficit-minded gang at Treasury: "Treasury folks are just happy to have a marker down on fiscal policy. We've been taking a lot of heat from the bond buyers who have been bitching and moaning that we're not serious on the debt, and this allows our team to push back."

Progressives—who days earlier had been down in the dumps about Obama's budget-cutting deal with the GOP—were heartened. Author/academic George Lakoff praised the speech as a "work of art" and a return to Obama's "moral vision." Krugman (sort of) hailed the president: "Obama made the moral as well as practical case." Robert Greenstein, the head of the Center on Budget and Policy Priorities, praised the plan, though he noted the two-to-one cuts-to-revenues ratio was not sufficiently balanced and that Obama's call for $360 billion in mandatory program cuts could unintentionally lead to "substantial cuts in core programs for low-income Americans."

The establishment's budget hawks tended to praise Obama. Pete Peterson, the doyen of this crowd, proclaimed: "President Obama's proposed framework is a big step toward the compromise we need to achieve fiscal sustainability."

Obama's top advisers were pleased. The president had won over Washington's deficit-über-alles crowd and the progressives. After weeks of budget bruises, Obama seemed to be roaring back.

TWO DAYS AFTER OBAMA'S SPEECH, THE HOUSE REPUBLICANS took a leap of faith and approved Ryan's budget on a party-line vote.

Democratic strategists were giddy. This, they believed, was

the path back to power. And that day, Kathy Hochul, a Democrat running in a special election to fill an open congressional seat in a Republican district northeast of Buffalo, New York, began decrying the Ryan budget measure as part of her campaign against a Republican millionaire named Jane Corwin who supported Ryan's plan.

Soon lawn signs popped up in the district: SAVE MEDICARE/ VOTE HOCHUL; and the Democrats were running ads warning the Corwin-endorsed legislation would "essentially end Medicare." (A month later, Hochul won this open seat 48 to 42 percent.)

Across the country, Republican House members holding town hall meetings—including Ryan—were met by angry constituents decrying the GOP plan to end Medicare as a guaranteed service. The DCCC released an ad showing seniors who had lost Medicare mowing lawns, selling lemonade, and performing as strippers to raise money for their health care. One survey found that 78 percent of Americans opposed Medicare reductions.

This was all the more delicious for Democrats because the Republicans had won back the House in part by attacking Obama and the Democrats for cutting Medicare. (Obama had not actually cut benefits for Medicare beneficiaries; he had reduced payments for Medicare Advantage, a program that subsidized private insurance plans that operate at higher costs than traditional Medicare.) Now it was the Democrats' turn to stick it to the Republicans. In private, Pelosi repeatedly urged Obama to advance this attack on Republicans.

But instead the White House throttled back. The bashing of Ryan and the Republicans—over values, visions, and budget numbers—ceased. Within the White House, the order came down from on high to tone down rhetoric and not demonize the Republicans: we're going to need to make a deal on the debt ceiling—and soon.

Asked at the time why the assault evaporated, a top White House aide said, "After that speech, we moved to the process"—meaning the talks with the Republicans.

Some in the White House, though, were in a more pugilistic mood. Austan Goolsbee produced a whiteboard video explaining what was wrong and excessive about the Ryan budget. The White House decided not to release it.

From the outside, it looked as if a switch at 1600 Pennsylvania Avenue had been turned off. A former top Obama White House official was puzzled by the lack of follow-up to the anti-Ryan speech: "Why not say that five times? Emphasize why we won't cut off the vulnerable, why government investment is important. Why not follow up over and over?"

An outside Obama adviser also was taken aback: "There was forty-eight hours of knocking the shit out of Ryan. Why nothing after that?"

Axelrod subsequently explained, "The decision was made to go out and talk about jobs and the economy and allow the negotiations to proceed until the president needed to intervene." And as Robert Gibbs put it, "It's difficult to put out your right hand to shake their hands and then strike them with your left hand."

Democrats and progressives questioned the back-to-talks strategy and yearned for a political brawl reminiscent of the Clinton-Gingrich clash of the 1990s. This political dreaming rankled some within Obama's circle. It's easy to criticize and concoct combative political strategies, White House aides complained, when it's not your job to avert economic calamity.

The president wasn't doing what he *wanted* to do; he was doing what he believed he had to do. "The president is not wild about his message now," a top Obama adviser said at the time. "And he understands that if independents won't listen to him, he can't bring them to where he needs to bring them. He is anxious to get to the next chapter."

—⁓—

WHAT WAS MOST CONCENTRATING THE COLLECTIVE MIND OF the White House was the debt ceiling. In an April 4 letter to Reid and Boehner, Geithner had reported that the US government would hit its limit on borrowing in mid-May—though Treasury could take steps to delay the moment of reckoning for weeks.

There was no time to waste.

Obama and his top economic advisers—Geithner, Sperling, Lew, and others—felt as if they were dealing with a ticking time bomb. (Later the default D-day would be pushed back to August 2.) They had taken their shots at Ryan and now needed to move on to negotiations that had to succeed.

Pondering the criticism from the left, Sperling thought of an old joke. A physicist, a chemist, and an economist are stranded on a desert island. One day, a can of tuna fish washes ashore. But they have no way to open it. The three decide to take twenty-four hours for each to come up with a possible solution.

The next day, the chemist tells the others he has found several chemical compounds he can place at the edge of the can. Once exposed to sunlight, they might burn a hole in the can. The physicist reports that he has found a boulder slanted on one side. If the can is dropped on this part of the boulder at a certain angle, it might break open. Then it's the economist's turn. He says, "Now suppose we had a can opener. . . ."

Sperling saw the critics as that economist. They wanted to relax the one constraint that could not be relaxed: Republican intransigence. But Obama and his crew were the responsible party; if they didn't drive a resolution with the Republicans, the economy could crash. They couldn't assume away the stakes. Wishing for a can opener wouldn't help.

Was this overly cautious? Overly deferential to political reality? Obama and his aides were outraged that the Republicans were using the possible default of the United States as a bargaining chip. Obama wanted a clean debt-ceiling hike, no strings (or cuts) attached—though such a bill wouldn't likely pass the House and pushing it would leave Obama open to the charge (false as it was) that he was in favor of more national debt, while the Republicans were championing fiscal discipline.

Obama and his team had decided there was little they could do to alter the basic dynamic at hand. Could they deny the hostage taking that had occurred? Could they call the GOP's bluff—dare Boehner not to raise the debt ceiling without spending cuts—and take a chance on default? It didn't help that the economy was once again in a precarious condition.

About this time, Goolsbee surveyed the gathering forces and thought of the Battle of Gettysburg. When he and his family had toured the Civil War site in nearby Pennsylvania, he had been impressed by a fundamental question: Why did this pivotal engagement of the Civil War take place at Gettysburg? There was not much there. The location did not have tremendous strategic importance.

But the great armies of the North and the South had been on the move and a collision was coming. The two stumbled upon each other in the small Pennsylvania town, and the inevitable and decisive battle ensued.

To Goolsbee, who would soon leave the White House and return to the University of Chicago, it looked as if a monumental clash between two opposing forces was now unavoidable and the confrontation over the debt ceiling would be the political and policy equivalent of Gettysburg: a big and bloody fight.

Chapter Ten

MISSION ACCOMPLISHED

EVERYONE WAS LAUGHING.

President Obama was standing at the podium in the main ballroom of the Washington Hilton, addressing the tuxedo-and-gown crowd at the White House Correspondents' Association dinner on the last Saturday night of April 2011.

The annual event is mockingly referred to as Washington's prom, and each year thousands of journalists, government officials, politicos, and a smattering of Hollywood celebrities gather to drink, eat, schmooze, observe, and be observed. Most years, the president attends and is compelled to perform like a stand-up comic. Obama had the assembled in stitches.

The president was targeting his best jabs at Donald Trump, the celebrity billionaire then flirting with joining the race for the Republican presidential nomination. Trump had leaped into the fray waving the banner of the birthers—those conservative Obama foes who claimed the president had actually been born in Kenya, rather than in Hawaii, thus making him ineligible to be president.

This outlandish charge had dogged the president since the

campaign—even though the state of Hawaii had issued a routine certification of birth proving that Obama had been born there. (Birth notices in two local newspapers from the time also confirmed this.) The birthers' conspiracy theory reflected a predominant theme within the conservative opposition: Obama was the other—somehow not quite a real American who truly understood and treasured the nation.

Birtherism had long lost steam, but Trump had revived the cause. He spent weeks repeatedly raising questions about Obama's birth and called on the president to make public his original long-form birth certificate. Sarah Palin endorsed his effort. Recent polling indicated that 45 percent of Republicans believed Obama had been born in another country.

Then on April 27, Obama strode into the White House press briefing room to announce he was releasing his long-form birth certificate, which showed—surprise, surprise—that he was Hawaiian born.

"We do not have time for this kind of silliness," Obama remarked. "We've got better stuff to do. I've got better stuff to do." And he took a sharp poke at Trump: "We're not going to be able to solve our problems if we get distracted by sideshows and carnival barkers."

Pundits immediately questioned whether Obama had finally relented to the birthers' call because the White House was worried that the issue, thanks to Trump, was gaining traction. White House officials' simpler explanation was that Obama had just had enough.

The week Obama gave his speech at George Washington University, challenging the Republicans on economic policy, he had been disheartened to see Trump and his birther claims draw so much attention. That's it, he thought. He instructed his attorney to contact Hawaii and request a copy of the long-form certificate. And Obama decided that he would make the official announcement before the television cameras.

"He wanted to deliver the overall message to the media and the public that we ought to focus on the real stuff," a White House aide said.

Trump declared he was "very proud of himself" for having compelled Obama to produce that document. (Some birthers would insist the certificate had been forged.) But Obama would literally have the last laugh.

At the dinner, three days later, the president was pummeling Trump—with gags.

"No one is happier, no one is prouder to put this birth certificate matter to rest than The Donald," Obama quipped. "And that's because he can finally get back to focusing on the issues that matter—like, did we fake the moon landing?"

Referring to an episode of Trump's reality television show, Obama cracked, "The men's cooking team did not impress the judges from Omaha Steaks, and there was a lot of blame to go around. But you, Mr. Trump, recognized that the real problem was a lack of leadership. And so ultimately, you didn't blame Lil' Jon or Meat Loaf. You fired Gary Busey. And these are the kind of decisions that would keep me up at night. Well handled, sir."

The audience roared with laughter. Trump was at the dinner, a guest of the *Washington Post,* and practically everyone in the room was looking to see his reaction. He didn't crack a smile. He appeared to be seething. There was a charge in the ballroom. It was as if a school bully were being called out in the middle of an assembly. In this instant, Obama was vanquishing Trump, birtherism, the silliness that often infects American politics, and the far Right's ridiculous assaults on him.

In a day, though, this moment would seem sharper and deeper than the audience at the dinner realized, for as Obama was mocking Trump, he was anxiously awaiting the outcome of one of the most weighty decisions of his presidency—his order to send a team of US special forces into Pakistan to get Osama bin Laden.

OBAMA HAD MADE THE FINAL DECISION TO LAUNCH THIS DANgerous and daring mission the day before. And as he joked that Saturday night, the president and several aides—only a small number had been informed about the raid—realized the next day could mark the virtual end of Obama's presidency if the operation went badly.

That Saturday morning, Denis McDonough brought an extra suit to work, just in case he'd be stuck at the White House for a few days. Valerie Jarrett learned of the possible raid only that day. She was nervous. She had wanted to stay home all day, but she forced herself to go to the correspondents' dinner and the various parties and receptions before and after.

She remembered that a few days earlier she had been with the president in New York City for a fund-raiser. He was working the rope line, and a woman introduced herself, noting in a near whisper that she had lost a relative in the 9/11 attacks.

"Are we going to get bin Laden?" the woman asked.

Holding her hand, Obama replied, "We're working on it."

At the dinner, Bill Daley was sitting at the ABC News table. Next to him was Eric Stonestreet, a star on the network's hit sitcom *Modern Family*. Stonestreet was excited because he was booked on a White House tour the following day.

At one point, Stonestreet checked his e-mail, and his face registered dejection.

"Oh shit," he muttered. The tour was canceled. He looked to Daley for an explanation. Daley didn't tell the actor that he was the one who had called off the tours for the day. He didn't want tourists in the White House at a time of potential crisis.

"Maybe a pipe burst," Daley said.

During the 2008 campaign, Obama had bluntly vowed, "We will kill bin Laden." In the spring of 2009, the president wondered if the hunt had slipped off the front burner.

Consequently, in June 2009, he sent a memo to Leon Panetta, the new CIA chief, stating that he considered finding bin Laden a high-priority task. He requested a detailed operation plan for locating and "bringing to justice" the al-Qaeda leader.

In the months that followed, Obama would receive an occasional briefing on the search for bin Laden. But there was never any information of consequence—until the summer of 2010. The president was told about a compound of interest in Abbottabad, Pakistan, about thirty-five miles north of Islamabad, a suburban and affluent area that was home to a Pakistani military academy. OBL might be there.

In the years after the 9/11 attacks, US military and intelligence officials interrogating suspected terrorist detainees continuously sought information on individuals who might be assisting bin Laden. As a result, they learned of a courier who was said to be trusted by bin Laden and who had been a protégé of Khalid Sheikh Mohammed. But all they had was the man's nom de guerre.

For years, the CIA tried—and failed—to determine his true identity and location. Then in 2007, the CIA discovered his name—Abu Ahmed al-Kuwaiti—and two years later they identified areas in Pakistan where he and his brother operated. But both men utilized extensive security measures to cloak their actions and whereabouts; this intensified the CIA's interest in them but made it harder to find either of the two.

In the summer of 2010, US intelligence intercepted a telephone call between al-Kuwaiti and another operative being watched by the CIA. They located al-Kuwaiti and his brother and tracked them to the Abbottabad compound.

Intelligence analysts were stunned when they received details of the residence. It was much larger than nearby homes, and there were extensive security features. The main building had few windows facing outside, and it was ringed by twelve- to eighteen-foot security walls topped with barbed wire. Residents of the compound burned their trash rather than placing it outside for collection. It seemed to have no phone or Internet service. The two brothers, who were living there, had no source of wealth. The place appeared custom-made for a "high-value target."

And a family—matching a description of the bin Ladens— was living within the compound. The analysts concluded there was a good chance this was bin Laden's home.

In a meeting so secret that White House aides did not e-mail about its subject, Panetta informed Obama, Joe Biden, Robert Gates, and Hillary Clinton about the compound, and he showed the president an image of the residence. The atmosphere was electric. Finally, after all these years, they had a bead on bin Laden.

At this point, Obama couldn't do much more than wait for the spies to do their jobs and gather more intelligence. But the president did start thinking about options. During the presidential campaign, he had repeatedly pledged that he would move unilaterally, if necessary, against "high-value terrorist targets like bin Laden" in Pakistan.

Hillary Clinton had chastised him for talking recklessly. John McCain had derided Obama for announcing that "he's going to attack Pakistan." Now, Obama was close to being able to make good on his promise.

The CIA continued to pursue more intelligence, looking for incontrovertible evidence. Satellites snapped high-resolution photographs of the compound; eavesdropping equipment was trained on it in an attempt to record voices inside.

From the start, a paramount concern for the White House and the CIA was security. No one wanted the slightest word of this to leak. Just a handful of Obama's national security aides were informed about the lead on bin Laden.

Obama and his top national security advisers feared that if even the most vague reference reached the media—say, a newspaper reported that the CIA had stepped up the hunt for bin Laden—a FOR SALE sign would immediately appear outside the Abbottabad compound. Obama would not even discuss the operation at his biweekly counterterrorism meetings with cabinet officials.

The CIA, working with other intelligence agencies (including the National Security Agency and the National Geospatial-Intelligence Agency), continued to gather intelligence on the compound, and even as the agency's analysts became increasingly convinced that bin Laden was there, they always had doubts. Perhaps it was the home of a Dubai prince lying low—or Afghan Taliban chief Mullah Omar, or al-Qaeda number two Ayman al-Zawahiri.

The president and a few of his aides tracked the CIA's work, and by mid-February 2011, they concluded the intelligence was strong enough to start planning a mission.

Panetta brought Vice Admiral William McRaven, the commander of the military's Joint Special Operations Command, to

CIA headquarters to tell him about the compound. McRaven was the military's top expert in covert warfare. He and his senior aides at JSOC immediately began cooking up various options for a secret assault.

Over the course of several weeks, McRaven developed three basic COAs—courses of action. A massive bombing strike in which B-2 stealth bombers would drop several dozen two-thousand-pound GPS-guided bombs and obliterate the entire compound; a helicopter raid mounted by US commandos; or a joint assault with Pakistani forces, who would be informed of the operation only shortly before its launch.

It was getting real. As Obama contended with the challenges of the Arab Spring, the frustrations of budget negotiations with the Republicans, and the barely improving economy, he was thinking about achieving the biggest victory in the so-called war on terrorism (a phrase his administration rarely used). He could be the president to take down the nation's number one enemy.

IN MID-MARCH 2011, OBAMA CONVENED A SERIES OF NATIONAL Security Council meetings on the bin Laden operation. At the first, on March 14, Panetta presented Obama with the three COAs. According to a participant, the president had "a visceral reaction" against this bombing strike because collateral damage would likely extend beyond the compound into the surrounding neighborhood. The CIA had already determined that the compound contained a number of women and children.

Another drawback of such an attack was that it would leave behind only rubble—and the remains of twenty or so people mixed in with the concrete and steel. In all that wreckage, could they find a piece of bin Laden—hair or flesh—for DNA analysis?

"The question was, would you accrue the strategic benefits of getting bin Laden if you couldn't prove it?" Ben Rhodes recalled.

It could take several months for the intelligence community—by monitoring chatter in the al-Qaeda network or other means—to confirm bin Laden's demise.

"And what could be worse," Nicholas Rasmussen, the White House senior director for counterterrorism, later noted, "than OBL survives and comes out and says, 'The United States failed to kill me'?"

Obama all but scratched this option off the list. He did ask the military to consider a surgical strike targeting the specific person living within the compound whom intelligence analysts suspected was the al-Qaeda leader. The analysts had dubbed him "the pacer," for he would stride around the compound as if for exercise.

The CIA had not been able to obtain physical evidence that confirmed the man was bin Laden. But the National Geospatial-Intelligence Agency, which analyzes overhead imagery, had determined this fellow was between five foot eight and six foot eight. Far from definitive. And he might be a decoy. But if they targeted him with a drone missile strike, would they really know it was bin Laden?

The president appeared to be leaning toward an assault. He pressed McRaven on the details. How much training time would a squad need? How could the commandos make a positive identification of bin Laden?

Gates and others, though, were skeptical—or, as one participant later recalled, "keenly focused on the risks associated with a helicopter raid." Gates had been at the CIA during Desert One (the failed mission President Jimmy Carter had approved to rescue American hostages held in Tehran in 1980) and Black Hawk Down (the 1993 operation in Somalia that fell apart and resulted in the deaths of eighteen US troops). Another high-risk helicopter raid—this time flying undetected a hundred miles into another country—was worrisome to him.

Obama and his aides dismissed a joint raid with the Pakistanis. They simply could not trust them. US officials had long suspected Pakistan's Inter-Services Intelligence of routinely tipping off targets of US actions, including drone strikes.

If bin Laden was indeed in the Abbottabad compound, it was hard to believe that no Pakistani military or intelligence officials knew he was there. Perhaps there was a way to work with the Pakistanis: if Washington on the morning of a planned attack contacted Pakistan's top military chief, General Ashfaq Parvez Kayani, and essentially forced him to join the raid on a moment's notice. Still, Obama did not believe any other nation, let alone Pakistan, could be trusted with advance information of an assault.

Soon afterward, officials reported back to the president that a surgical strike would require a large amount of ordnance. That is, it wouldn't be all that surgical. A helicopter raid appeared the only viable choice.

As the planning meetings proceeded—the president and his aides often had a model of the compound before them—a critical point about a unilateral US assault caught Obama's attention: How would these covert warriors return safely from the compound, especially if they were to encounter hostile Pakistani military forces? He noticed that in the initial planning the assault force was small. He asked McRaven if such a force could fight its way out if necessary.

McRaven had based the planning on an assumption that if his commandos were confronted by the Pakistanis, they would protect themselves without attempting to defeat the Pakistani forces, while waiting for the politicians in Washington and Islamabad to sort things out. He calculated that his team could hold off any Pakistani assault for one or two hours.

Obama nixed the idea of commandos hunkering down to await diplomatic rescue. He worried that the Navy SEALs conducting the mission could end up as hostages of the Pakistanis, and he told McRaven to ensure that the US forces could escape the compound and return to safety, whether or not they encountered Pakistani resistance.

"Don't worry about keeping things calm with Pakistan," Obama said to McRaven. "Worry about getting out."

McRaven added additional forces; a second group of SEALs would be prepared to take on any Pakistani forces that might try to intervene.

The pace of the meetings in the White House—still involving merely a few aides—and at the CIA and JSOC picked up.

Throughout April a team of Navy SEALs from the elite and supersecret Team 6 unit (which was developed in response to the Desert One debacle) practiced for the assault, using a site built to resemble the compound. From a safe house it had set up earlier in Abbottabad, the CIA continued to watch the compound.

ON THE EVENING OF THURSDAY, APRIL 28, SHORTLY AFTER ANnouncing that Panetta would succeed Gates as defense secretary, the president convened what would be the last NSC planning meeting on the bin Laden operation. McRaven was already in Afghanistan, ready to launch the raid if so ordered. Security concerns prevented him from being looped into the meeting via video.

Mike Mullen described the latest plans. In response to Obama's demand for a fight-your-way-out plan, McRaven had added two Chinook helicopters to complement the pair of MH-60 Black Hawk stealth copters. And the SEALs had been instructed to be prepared to engage if they were met by the Pakistani military.

Panetta reported that the intelligence community had conducted a red team test exercise, in which analysts who had not previously worked on the bin Laden case evaluated the intelligence that had been gathered. The CIA had earlier told Obama that its analysts had concluded there was a 60 to 80 percent certainty that bin Laden was in the Abbottabad compound. The red team ended up with lower odds: 40 to 60 percent.

Several of Obama's national security advisers were worried by the red team results and wondered why the confidence level was lower. Michael Leiter, the chief of the National Counterterrorism Center, believed the CIA had inflated the case. But the president showed little interest in calibrating the difference between the two conclusions.

"My impression was that he viewed this as a fifty-fifty case," Rhodes later said. "He knew there would always be a significant chance that the intelligence was not correct and he would have to live with the downside risk."

Whatever the percentages, Obama and everyone else realized this was the best shot the US government had at finding bin Laden since he escaped at the battle of Tora Bora in late 2001.

The group reviewed the three COAs now on the table: a unilateral US raid, a missile strike from "a standoff position" (probably using a drone), or postponing any action while more information was collected.

Obama and his aides went over the negative scenarios. Bin Laden might not be there; he might escape the commandos. The compound could be rigged to explode if attacked. The mission could end in a firefight with the Pakistanis, and that could ignite anti-US demonstrations throughout the country and cause a breach in the already delicate relationship with Islamabad.

Neither Obama nor anyone else in the room had to discuss the most obvious point: Obama would be placing his presidency on the line. If he ordered the mission to proceed and it succeeded, he would receive hurrahs. It could well boost his political standing.

Yet if the mission went sour—and American lives were lost, and the United States embarrassed—Obama would be denounced as hapless and incompetent. His right-wing critics would have evidence for their otherwise baseless contention that he was an amateur on national security. Obama and his aides were well aware that Desert One had reinforced the charge that President Carter, then contending with economic troubles, was feckless, and that the failed mission severely damaged his reelection prospects.

It was entirely possible—arguably probable—that Obama might never recover from such a failure. So much had to go exactly right for the raid to succeed; one accident or screwup (well beyond Obama's control) could sink the operation—and his presidency. Obama was on the verge of taking the ultimate political risk.

It was time to decide. Obama asked his national security team for their individual recommendations. Biden and Gates each counseled waiting for more intelligence. "Don't go," the vice president said. Several of the participants still opted for a missile strike. Panetta and the president's chief counterterrorism adviser, John Brennan, backed the helicopter assault. There was no consensus.

The raid was not supported by a majority—and Obama's most experienced advisers, his vice president and defense secretary, were wary of proceeding with this mission.

"He had very senior people advising a different course," a participant recalled. "It would take a lot of confidence and fortitude to go against all that."

At the end of the meeting, Obama said, "I'm not going to tell you what my decision is now. I'm going back and thinking about it some more." But he said he would decide soon.

Obama left the meeting, walked across the colonnade past the Rose Garden to the residence, to make a decision. He was thinking of Desert One and Black Hawk Down.

Obama, despite the public perception that he was a cautious seeker of consensus, was a risk taker. He had pushed for health care reform when practically no one in the White House thought it could be achieved. He had advocated pressuring Mubarak, though his top foreign policy aides preferred preserving the status quo. He had launched a military action in Libya over the reservations of his most experienced military advisers. Yet again, the president was considering disregarding the play-it-safe advice of senior national security advisers to move forward with a bold action that had no guarantee of success and that could bring about a heap of trouble.

Dealing with Congress had forced Obama into the role of a hostage negotiator who believed that he had to pursue compromise (perhaps not to his liking) to prevent economic harm. As commander in chief confronting challenging circumstances, he remained deliberative, but he also had the latitude to be decisive and daring.

ON FRIDAY MORNING, BEFORE FLYING TO ALABAMA TO INSPECT tornado damage, Obama called Donilon, Daley, McDonough, and Brennan to the Diplomatic Room of the White House. They had expected to brief Obama again on the bin Laden options. But before the aides proceeded, Obama interrupted: "It's a go."

"I've never seen a decision that courageous done without any sense of drama," a top national security aide subsequently observed.

Obama's orders were conveyed to McRaven. The admiral now had to choose when to launch the raid. He had briefed the president that the next five or six days, starting that Saturday, were the optimal time frame, but the sooner the better.

Now all Obama and his aides could do was wait.

"Those last few days," a senior administration official recalled, "we were all terrified that we'd come in for the latest meeting and Mike Morell [the deputy CIA director] would say the compound is empty."

On Saturday, McRaven opted against launching the mission, due to bad weather. (Had it not been for this hiccup, news of the raid might have emerged during the correspondents' dinner.) Later that day, while on a break from rehearsing for the dinner that night, Obama called McRaven. He asked if he had learned anything while in Afghanistan that had altered his confidence level in the mission or its risk profile. Nothing has changed, McRaven replied. The raid was likely to occur tomorrow.

"There's no one I'd rather have doing this than you guys," Obama told him. "Godspeed."

On Sunday morning, May 1, Obama played golf at Andrews Air Force Base. He returned to the White House at 2:00 P.M. By then, his top national security advisers were assembled in the Situation Room. Panetta, who was in the CIA command center, was connected via a video link.

Panetta reviewed for the president, vice president, and the others what was about to happen. The helicopters, carrying seventy-nine commandos, were on their way from a US base in Jalalabad, Afghanistan. It was a moonless night. At about 3:00 P.M., Panetta announced, "They've crossed into Pakistan."

Panetta was being briefed by McRaven, who was commanding the operation from Jalalabad. The admiral was providing intermittent updates to Panetta, informing the CIA chief of the milestones being met. It was not a running commentary. Ten or fifteen minutes could go by with no new reports.

"The minutes passed like days," Brennan later said.

Obama could only recall one other moment in his life more nerve-racking: when his daughter Sasha, then three months old, contracted meningitis, and he had waited for a doctor to tell him if she would live.

In a small breakout room next to the Sit Room, Brigadier General Marshall "Brad" Webb, the assistant commander of JSOC, was monitoring another feed that was providing real-time reporting.

"You could see some of what was going on," a senior administration official recalled.

This feed had not been patched into the Situation Room; officials did not want it to seem that the president and his advisers were micromanaging the actual operation.

"There was a decision to keep the blow-by-blow separate," this official said. "But do you want to hear or *see* the movie?"

Eventually, everyone crowded into the small room to watch and listen with Webb. It was a tense scene that would be captured in an iconic photograph showing Obama, Biden, Gates, Clinton, Mullen, Daley, McDonough, Brennan, and others staring intently at a video feed.

After the mission, White House officials would not identify what the president and his advisers had been watching at that moment in time.

"No one knew everything that was going on inside the compound," an administration aide recalled. "There was just a big-picture strategic awareness."

As the mission proceeded, Panetta relayed the information he was receiving from McRaven. On a video screen, there were night-vision images from a drone.

The raid nearly failed before it began; a helicopter carrying Navy SEALs stalled as it approached the target and tumbled to the ground. The pilot managed to land it, and no one was hurt. The mission could continue. As Obama watched and listened, he was calm and quiet. Biden fingered his rosary beads.

The SEALs moved in, and a short firefight broke out. The commandos killed al-Kuwaiti and his brother, entered the main building, made their way to the third floor, and located bin Laden. One of bin Laden's wives rushed the commandos and was shot in the leg but not killed. Bin Laden ran from the SEALs, and they shot him in the head and killed him.

Panetta narrated: "They've reached the target. . . . We have a visual on Geronimo [the code name for bin Laden]. . . . Geronimo EKIA."

Enemy Killed in Action.

For a moment, there was silence at the White House; neither the president nor his aides uttered a word. There was no fist pumping, no cheering.

"The mood was one of sober relief," a senior administration official recalled. "There was no joy in this."

The president simply said, "We got him."

The commandos grabbed computers, hard drives, flash drives, and documents. The target, the SEALs believed, was indeed bin Laden; one of his wives and several children living in the compound confirmed this.

The SEALs had to blow up the disabled helicopter to keep it out of unfriendly hands and departed in one of the original assault helicopters and one of the two backup copters that had been added to the mission in response to Obama's instructions.

They were out within thirty-eight minutes, with the body of bin Laden—and heading to USS *Carl Vinson* in the North Arabian Sea. There were no American casualties.

But the nail-biting continued as Obama and his team waited

for the Navy SEALs to clear Pakistani airspace. Only after the American forces had done so did Obama call Pakistani president Asif Ali Zardari to inform him of the raid.

Photos of bin Laden's dead body arrived at the White House and were passed around the table in the Situation Room.

"We were not experts," one senior administration official later said. "But it looked like him."

At one point, a commando had lain down next to the body so the SEALs could get a sense of the man's height; the SEALs had neglected to bring a measuring tape. The president joked, "We donated a $60 million helicopter to this operation. Could we not afford to buy a tape measure?"

About three hours after the commandos had left Abbottabad, Obama was told there was a "high probability" the dead man was bin Laden. A DNA sample had already been sent electronically to the United States, where scientists would within a day match it against DNA from bin Laden family members.

Within hours of the assault, bin Laden's body was washed and placed in a white sheet, in conformance with Islamic practices, and buried at sea. The most expensive and expansive manhunt in US history, perhaps in the history of the world, was done—mission accomplished.

ABOUT 9:45 P.M., DAN PFEIFFER SENT OUT A TWEET: "POTUS to address the nation tonight at 10:30 PM Eastern Time."

Obama's first instinct was against rushing out a public statement until the DNA evidence was confirmed. But his aides had persuaded him he couldn't sit on such a huge story.

Reacting to Pfeiffer's tweet, media outlets immediately began speculating; some journalists guessed that bin Laden had been found.

As White House aides were leaving the Situation Room to work on the president's statement, one noticed a television showing *Celebrity Apprentice*. They chuckled over the timing: soon the news of the killing of bin Laden would leak and, no doubt, knock Trump's reality show off the air.

While working with Rhodes and other aides on his statement, Obama pointed out a significant aspect of the raid.

"It's important to be able to prove that America could do something difficult that takes a long time," he said.

The government, the president reflected, had pursued a complicated thread to its end and finished the job by taking a risk.

At 10:25 P.M., the chief of staff for former Defense secretary Donald Rumsfeld tweeted, "I'm told by a reputable person that they have killed Osama Bin Laden."

The news zipped across the Internet. NBC, ABC, and CBS interrupted their programming to report the news.

At 11:35 P.M., Obama entered the East Room, where more than a dozen aides were waiting for him, including Biden, Daley, Clinton, Panetta, Mullen, and Donilon. Speaking to the nation—and the world—in a televised broadcast, he announced that the United States had killed bin Laden. He offered a brief and vague description of the raid.

"Our war is not against Islam," Obama stated.

He did what he could to smooth over any problems the assault caused with Pakistan. "Our counterterrorism cooperation with Pakistan helped lead us to bin Laden."

Obama asked Americans to "think back to the sense of unity that prevailed on 9/11. I know that it has, at times, frayed. Yet today's achievement is a testament to the greatness of our country and the determination of the American people."

When the short speech was over, he shook hands with several aides. There were a few backslaps, not too many. Then the president walked through the colonnade. A flag-waving crowd, many of them college students, had gathered in front of the White House.

The president could hear them cheering: "Obama got Osama."

THE RAID WAS OVER. BIN LADEN, THE MAN WHO HAD ORCHES-trated the massacre of thousands of Americans, who had forced changes throughout US society, and who had pulled the United States into two wars, was dead. Now there would be the inevitable media frenzy, as the nation processed the event.

Journalists would demand details of the assault. Questions would materialize. What did the Pakistani government know of the compound and bin Laden's presence there? Had the United States tried to capture bin Laden or were Obama's orders to kill him on the spot? Should evidence of bin Laden's death—such as the photographs—be publicly released? Was the intelligence that led to al-Kuwaiti a result of the so-called enhanced interrogation techniques—or torture—used during the Bush-Cheney years?

The next day, Jay Carney brought Brennan to the daily briefing to answer a few of the emerging questions. The commandos had been prepared to take bin Laden alive had he presented no threat and surrendered, Brennan said, adding that this had been a "remote" possibility. (Various media reports, quoting unnamed military officials, said the SEALs knew all along that their mission was—nod, nod, wink, wink—not to capture the al-Qaeda leader.)

As for Pakistan, Brennan reported that the administration was in contact with officials in Islamabad to determine who in Pakistan had been assisting bin Laden. He declined to speculate on whether anyone within the Pakistani government might have been aware of bin Laden's presence in Abbottabad. Brennan told the press corps that there was yet no decision on whether to release photos of bin Laden's corpse—reportedly gruesome and bloody images.

Brennan provided a more detailed account of bin Laden's final moments: he had been "engaged" in the firefight when he was shot and he had tried to use a woman, presumed to be his wife, as a shield, and this woman was killed. The next day, Carney would have to correct the record: bin Laden had not been armed, but he had resisted capture. His wife had been shot in the leg, not killed, while rushing a commando. Bin Laden had not used her as a human shield. Subsequently, the official account changed again: there had only been an exchange of fire when the SEALs first landed, no later firefight involving bin Laden.

The shifting story, for the moment, tainted the achievement. One former CIA official grumbled at the time, "A near flawless intelligence and special ops mission turned into a political circus discussion."

White House officials dismissed complaints, noting they had tried to release as much information as they quickly could. One reason for the prompt dissemination of details was to be certain that the distrusted Pakistani intelligence service did not put out its own version of the events first.

In the raid's aftermath, a more significant scuffle transpired over the role of enhanced interrogation techniques in the CIA's hunt for bin Laden. In his first full day as president, Obama had signed an executive order banning torture. Ever since then alumni and allies of the Bush-Cheney administration had been arguing that the order was misguided and that the use of waterboarding and other harsh tactics was legitimate, necessary, and productive. Now this band seized on the bin Laden raid for justification.

John Yoo, a Bush Justice Department official who wrote secret legal memos justifying brutal interrogations, claimed that had Obama been running the show in the years after 9/11, "there would have been no enhanced interrogation program, no terrorist surveillance program, and hence no intelligence mosaic that could have given us the information that produced this success."

Yoo was basing his argument on the phrase "could have."

Representative Peter King, a New York Republican, was more direct: "We obtained that information through waterboarding. So for those who say that waterboarding doesn't work, who say it should be stopped and never used again, we got vital information which directly led us to bin Laden."

Carney insisted otherwise: "No single piece of information led to the successful mission that occurred on Sunday, and multiple detainees provided insights into the networks of people who might have been close to bin Laden."

Khalid Sheikh Mohammed had been waterboarded 183 times and had not provided accurate information about al-Kuwaiti, the courier, to his interrogators. (The Associated Press reported that KSM did offer up nicknames of couriers—but under standard interrogation, long after the waterboarding.)

Republican Senator Lindsey Graham, no ally of the White House, challenged the effort of torture-friendly conservatives to exploit the raid: "This idea we caught bin Laden because of waterboarding I think it's a misstatement. . . . This whole concept of how we caught bin Laden is a lot of work over time by different people and putting a puzzle together. I do not believe this is a time to celebrate waterboarding."

Three days after the raid, the White House announced there would be no photos released of bin Laden's corpse. Obama had asked his advisers to give him one good reason to release the photographs. Clinton consulted with Middle East allies, and they said a release was not in their interests. Gates fretted that the public display of bin Laden's dead body could trigger attacks on US military bases and embassies.

Obama explained his decision not to release the photos during an interview with *60 Minutes:* "It is important to make sure that very graphic photos of somebody who was shot in the head are not floating around as an incitement to additional violence. . . . We don't need to spike the football."

Obama did take a muted victory lap. On May 5, he went to New York City and joined 9/11 families in laying a wreath at Ground Zero; then he met with firefighters and police officers who had lost comrades in the attack on the World Trade Center. The next day, he flew to the army base at Fort Campbell in Kentucky to thank the SEALs who had conducted the OBL raid.

Afterward at a rally, he declared, "We're still the America that does the hard things." Thousands of soldiers cheered.

THE BIN LADEN RAID WAS NOT A ONE-OFF; IN A WAY, IT REPRE-sented a shift in foreign policy. Obama was determined to view the fight against Islamic extremism not as a geostrategic conflict that required, say, the invasion of another country, but as a battle against specific targets: al-Qaeda and its affiliates.

"It's an identifiable organization," one national security aide said, explaining the president's view, "and we can take out members. The focus is on actual terrorists."

After the raid, Obama's foes found it difficult to claim he was not a vigilant and forceful commander in chief. Since the campaign, Republicans and conservatives had derided Obama as a foreign policy weakling. In 2008, Sarah Palin had accused Obama of "palling around with terrorists." After Obama moved into the White House, Dick Cheney seemed to rise out of retirement every few months to accuse Obama of imperiling the nation. One time he harrumphed that Obama and his aides were "more concerned about reading the rights to an al-Qaeda terrorist than they are with protecting the United States."

In February, Newt Gingrich had slammed the Obama administration as "a very weak government led by a group of amateurs in foreign policy." Months earlier—in a birtherism-tinged remark—he had blasted Obama for possessing a "Kenyan, anticolonial" worldview and "being fundamentally out of touch with the way the world works."

The anti-Obama partisans who claimed that the president was not leadership material or that he was insufficiently concerned about threats to the United States (or that he was a secret, Kenyan-born Muslim socialist yearning to destroy the United States in order to gain dictatorial powers) would have a tough time selling that after the bin Laden raid.

The raid was also a good answer to the more measured criticism of the president.

"Before this, people might say he's not decisive, that he's detached and needs to be more in there," an outside Obama adviser remarked. "Now they see him as patient, thoughtful, thorough. This event reorients the process through which people will see him."

In a column, E. J. Dionne of the *Washington Post* observed that Obama "has now proved that he can be bold at an operational level."

When Jarrett read that, she huffed. Obama, as she saw it, was "a cool customer." She did not think this ought to come as a surprise to Dionne—or anyone else.

Polls conducted immediately after the raid showed a nine- to eleven-point bounce in Obama's approval ratings. But with unemployment still higher than 9 percent, gas prices hitting $4 a gallon, and Washington heading toward a debt-ceiling crisis, this would be a temporary lift.

"I was not that delighted after the bin Laden raid when the polls improved," the often-gloomy David Axelrod recalled. "I said to Plouffe the next step is predictable. There's bad economic news, gas prices go up, and there's a new narrative: 'Obama, who was ascendant a month ago, is now headed to disaster.'"

Within a week, the bin Laden bump in Obama's approval ratings indeed dissipated. Yet after several difficult months—which included the Egyptian and Libyan crises (during which Obama was frequently assailed as overly equivocal) and the ceaseless budget negotiations (during which Obama was often attacked for being disengaged)—the raid had cast Obama as a decisive commander in chief.

He had demonstrated to the American public that its government could succeed in an arduous task requiring patience, diligence, discipline, commitment, and boldness. There was, though, no telling how deeply that lesson would be absorbed. (In five months, at the end of September 2011, the CIA would succeed in killing Anwar al-Awlaki, the high-profile, American-born al-Qaeda leader in Yemen, with a missile strike, part of the expanded secret drone program authorized by Obama that had successfully targeted numerous al-Qaeda and Taliban commanders and their allies.)

Four days after the compound raid, Biden convened a group of bipartisan House and Senate members; their mission was to negotiate a deal on the deficit and the debt ceiling. What was to happen on this front would have more of an effect upon Obama's electoral prospects than his management of the masterful bin Laden operation.

As for Donald Trump, fifteen days after the bin Laden raid, he said he would not run for president. He broke the news at an event announcing that he would return with a new season of *Celebrity Apprentice* in the fall.

Chapter Eleven

RETURN OF THE
HOSTAGE TAKERS

"NOTHING IS AGREED TO UNTIL EVERYTHING IS AGREED TO," JOE Biden said—mimicking what John Boehner had repeatedly stated during the budget talks—as he met with a small group of lawmakers to open the debt talks that President Obama had requested in his speech at George Washington University a month earlier.

On May 5, 2011, at an oval table in the dining room of Blair House, a block away from the White House, Biden had assembled the Republican-designated negotiators, Eric Cantor and Jon Kyl; the Democratic team, Chris Van Hollen, Max Baucus, Representative Jim Clyburn, and Senator Daniel Inouye; as well as Timothy Geithner, Jack Lew, Gene Sperling, Jason Furman, and Bruce Reed.

Biden began by emphasizing the crucial principle that there's no deal until there is a deal.

With the president's blessing, Biden was hoping to create an environment in which anything could be put on the table—

a proposed spending cut, a possible tax increase—but nothing proffered or suggested was solid until a final agreement existed. If the not-until rule was in place, both sides would have more flexibility to toss out ideas and the discussions could be more candid.

The essential split at the table was no secret. The Republicans wouldn't raise the debt ceiling without deep spending cuts and budget reforms; the Democrats wouldn't agree to significant cuts (particularly any affecting Medicare or Social Security benefits) without an increase in revenues.

And Biden was firm on another key point: the deficit-reduction package they were trying to concoct could not be solely a cavalcade of cuts. The Democrats were indeed willing to concede on spending cuts to obtain a debt-ceiling hike, but they needed give from the Republicans. Biden was making sure that all the parties understood that there had to be revenues—either from closing tax loopholes for corporations or from gathering more taxes from those in the top income brackets. Neither Kyl nor Cantor raised an objection.

Let's proceed in good faith, Biden urged.

DAYS LATER, BOEHNER MADE THE JOB TOUGHER. IN A SPEECH, the speaker denounced government, claiming that many of the nation's woes were due to the "misguided belief by politicians that the American economy" can be "influenced positively by government intervention and borrowing." This was the anti-Obama view that government had no proactive role to play in bolstering the economy. More important, Boehner declared an ultimatum: he and his fellow House Republicans would not vote to increase the debt limit unless accompanying spending cuts were greater than the increase in the borrowing authority. Tax hikes, he asserted, would have to be "off the table."

Obama was facing what the administration viewed as

another hostage situation, this time with a ransom note demanding $2.4 trillion. Boehner and the Republicans were threatening to do worse than hold up middle-class tax cuts or shut down the government; they were threatening the default of the US government—and a possible financial meltdown.

The White House calculated that there was little chance of overcoming GOP opposition in the House without submitting to this demand; there were enough Republicans willing—if not eager—to shoot the hostage. Linking the debt ceiling to spending cuts, Obama and his aides believed, was dangerous. There had never been a connection in the past; this would establish a difficult precedent for the future. But Obama and his team didn't see much maneuvering room. The president would have to accept Boehner's basic terms.

THE BIDEN TALKS INITIALLY FOCUSED ON CUTS, NOT REVENUES, and the participants identified several hundreds of billions of dollars in reductions in "non-health mandatories"—such programs as farm subsidies and federal and military pensions.

Cantor appeared to have come prepared to bargain in good faith. He genuinely seemed to be seeking a package that could be approved by his caucus. There was almost a touch of optimism in the West Wing.

"Our own people would come back from negotiations and say the talks are going well, and we can't go out and bash the Republicans," a senior administration official recalled.

But there was a split between Cantor and Kyl. Cantor and Boehner favored a solution that would lift the debt ceiling for two years. They didn't want to deal with this again in six months, for they knew it would be a heavy lift to get the Tea Partiers and freshman members of their party to extend the government's borrowing authority even once, let alone two or three times before the 2012 elections.

Kyl seemed just fine with a short-term extension that would have to be revisited in six months or so. Mitch McConnell, who yearned to become Senate majority leader after the next election, didn't mind making Democratic senators facing reelection cast several uncomfortable votes for more debt.

The negotiators met five times during the month of May, eventually switching to a conference room in the Capitol. The initial sessions were productive. The group drew up a list of possible Medicare and Medicaid reductions that would squeeze hospitals, pharmaceutical manufacturers, and providers—not beneficiaries.

After the fifth meeting on May 24, Biden told reporters that he was confident they could reach at least $1 trillion in deficit reduction, as a down payment on a larger accord. There were only two problems: the Republicans had yet to indicate they could bend on revenues and $1 trillion was below Boehner's demand.

The Biden group seemed at least to be making more progress than the Gang of Six, a group of Republican and Democratic senators who had been working to craft their own broad-based deficit-reduction accord (with cuts *and* revenues). In mid-May, this effort seemingly collapsed, when Senator Tom Coburn, an Oklahoma Republican, withdrew from the gang, after Democrats wouldn't accept the deeper cuts he was pushing for Medicare.

During the Biden talks, tempers rarely flared, but at a policy conference in Washington at the end of the month, Sperling showed a touch of frustration with the overall Republican position.

"Serious people in serious discussion do not say you cannot have a single penny in revenues," he insisted.

Sperling noted that Representative Paul Ryan was imprisoned by a no-tax "ideological view." Without expanding revenues, Sperling asserted, "very severe cuts" were unavoidable. He pointed out that the cut in Medicaid proposed by the House Republicans would lead to millions of poor children losing health care and elderly Americans forgoing coverage for nursing home care.

"We are not criticizing their plan," he added. "We are just explaining their plan."

Yet Sperling still claimed he was upbeat about the Biden talks, remarking there was "a real seriousness of purpose in that room that gives me hope."

HOPE, THOUGH, WAS NOT BLOOMING ON CAPITOL HILL. ON June 1, a day after the Republican-controlled House voted down a bill to increase the debt ceiling (without any spending cuts), Obama had all the House Republicans to the White House for a not-so-intimate conversation.

The recent economic news had been lousy. The latest growth indicators were disappointing. But the topic was deficit reduction, entitlement reform, and the growing debt—an indicator of the narrow debate within the Beltway. The scores of GOPers sat in chairs in the East Room—almost as if they were there for a press conference—with the president at a table at the front.

Ryan was direct; he beseeched the president to stop "misdescribing" his Medicare plan and urged him to ratchet back on the "demagoguery." His fellow Republicans, who had taken a beating for supporting an end to the Medicare guarantee, gave Ryan a standing ovation for this heartfelt plea. Obama was fast to reply.

"I'm the death-panel-supporting, socialist, may-not-have-been-born-here president," he said, explaining that he had more interest than most politicians in eradicating extreme rhetoric.

What Ryan and the Republicans had experienced with

Medicare was nothing compared to the vitriol that had been hurled at Obama by their ideological soul mates. Obama held firm to his critique of the Republicans' Medicare plan: it was indeed a transfer of health care costs from the government to the elderly. The meeting accomplished little.

The next day, the House Democrats had their own turn at the White House. Many of them feared Obama was too willing to cave during the debt-ceiling negotiations. The president promised he would not again sign off on extending the Bush tax cuts for the top brackets, no matter what hostages the Republicans grabbed. Nancy Pelosi was adamant that Obama not trade away Medicare benefit cuts for a deficit/debt-ceiling deal.

Representative Henry Waxman, a veteran lawmaker from California, vented for the group. He urged the president to fight more fiercely and to take better advantage of the bully pulpit, implying that the president was an inadequate negotiator. This was the Democrats' number one complaint about the president—he gave away too much without enough of a fight. Many were still yearning for a more aggressive Obama.

The president insisted that he knew how to negotiate with the Republicans. He also told his party colleagues to stop expecting him to let loose the fiery rhetoric.

"When Eric Cantor says something, Eric Cantor says something," Obama explained. "When I say something, markets and countries and people react."

He was obliged to be more prudent. He did not have the luxury to engage in a political food fight.

Obama and his aides were frustrated that this seemed to escape their allies on the Hill, in the progressive activist community, and within the liberal commentariat.

Still, Obama said he believed he could reach an agreement on the debt ceiling within a month. Then they all could move on to talking about economic growth and jobs.

—᷒᷒᷒—

THE BIDEN TALKS ROLLED ON. CANTOR SENT AN E-MAIL TO House Republicans noting he was "cautiously optimistic" they would cobble up an agreement meeting Boehner's one-for-one demand. Then the negotiators moved to the tough stuff: revenues.

At a June session, Geithner presented a series of revenue proposals. He discussed closing loopholes and eliminating deductions. This included ending subsidies for oil and gas companies and tax breaks for corporate jets. All told, the Democrats were looking at an estimated $500 billion in added revenue as part of perhaps a $2 trillion deal.

"It was mainly cat-and-dog stuff, small stuff," said a Democratic staffer.

The Democrats could see Kyl and Cantor stiffening, as Geithner went through the revenue options, and the two Republicans responded without much enthusiasm.

"Their attitude," a White House official later recounted, "was, 'We'll let them *talk* about revenues. Raising the debt ceiling is enough of a concession for us.'"

A Democratic negotiator recalled, "We'd say, 'We know you don't want to do revenues, but which of these do you hate the least? How about corporate jets?' They didn't engage."

At one session, Kyl asked Lew and Sperling for clarification on a point that puzzled him: You say Medicare savings are good policy, but you will agree to them only if there are revenue increases? Yes, they said. Kyl scratched his head. But if the Medicare reductions are good policy, why not simply agree on this? The tax fight, he suggested, could be waged separately.

Lew and Sperling explained that the only way a package with Medicare savings could be acceptable to the Democrats was if it included revenues.

Kyl repeated his question: Why can't we go with what we can agree on? He seemed genuinely bewildered. Kyl didn't appear to realize he was asking them for total capitulation.

Throughout the talks, Biden was in touch with McConnell, who was urging the vice president to get Medicare into the mix of whatever deal his group was cooking up. Biden didn't have to guess why. If a bipartisan budget accord contained Medicare reductions, Republicans could say that everyone in Washington was for some sort of Medicare reform. The Democrats' most effective line of attack would be diluted.

Biden told McConnell to find an excuse to add revenues to the package.

AFTER THE BIDEN GANG'S NINTH MEETING ON JUNE 16, THE VICE president noted that crunch time was at hand. He told reporters that around-the-clock staff work and several meetings the following week would be necessary for his group to bang out that "real down payment" on a $4 trillion plan. But the final swaps, he noted, had yet to be made: "I'll trade you my bicycle for your golf clubs."

Yet this fiscal swap meet was becoming even more problematic because the Republicans were becoming more divided. Senate Republicans were saying a long-term debt deal looked unlikely and that Washington would have to settle for a short-term extension—that is, a repeat of this whole process later in the year.

Kyl publicly groused that the Biden talks had not generated sufficient cuts to warrant raising the debt ceiling at all. Asked by a reporter what sacred cows Republicans were willing to sacrifice to reach an agreement, Kyl sarcastically replied, "Gee, I can't think of any."

Cantor, though, was still hoping for an agreement that would avoid a replay in six months. "I don't see how multiple votes on a debt-ceiling increase can help us get to where we want to go," he said.

The White House was bargaining with a Republican Party at odds with itself.

IN WHAT WOULD BE THE LAST MEETING OF THE BIDEN GROUP, the Republicans pushed for Medicare reductions that would hit beneficiaries, not merely providers. The Democrats demanded closing tax loopholes and other revenue generators. The group had spent more than a month detailing significant cuts, but Kyl and Cantor had continued to resist a full conversation about revenues. Geithner reviewed revenue options.

"We have to get something," he said.

Van Hollen insisted that Medicare beneficiaries not take a hit—especially if corporations and well-off taxpayers were not paying more. The median income of a Medicare recipient, he pointed out, was under $23,000 a year. Why should Washington place greater financial burdens on these Americans than on oil companies, corporate jet owners, and hedge fund managers?

Cantor was not accepting any of the Democratic revenue proposals. But, he noted, the negotiators might be able to include revenues if they don't look like revenues. To the Democratic negotiators, he appeared to be seeking a sleight of hand with which he could fool the Tea Partiers.

Biden reminded the Republicans that ever since the opening minutes of these talks he had repeatedly stated that there could be no deal on any of the cuts without revenues.

Cantor replied that they still didn't have $2.4 trillion in cuts.

Biden pointed out that just about every serious deficit-reduction plan produced by think tanks, policy shops, and commissions included revenues.

Geithner was annoyed; the GOP side was demonstrating no flexibility.

Cantor responded, "We always said we have a vote problem"—meaning Boehner and Cantor couldn't sell a package with revenues to their caucus.

"The idea that we have to pay to avoid default—as secretary of the Treasury, I'm offended by that," Geithner said.

"We're running out of runway," Biden remarked.

Van Hollen had worried from the first meeting that during the talks Kyl and Cantor would pocket spending cuts the Democrats offered (provisionally!) and offer little in return. And that was now coming true.

"We were willing to take difficult measures in the context of a balanced agreement with shared sacrifice," a Democratic negotiator recalled. "We assumed we would have a revenue part."

Before the June 22 meeting concluded, Biden scheduled two more for later that week. His aim was to wrap up these talks and then kick the negotiations upstairs to Obama and Boehner. Only those two could ultimately decide how far to go on Medicare and what revenue proposals would be acceptable.

After the session, Biden told waiting reporters, "We're still moving."

Baucus stated, "There will be an agreement."

Following this meeting, Biden was chatting with Cantor and casually referred to a pending get-together between Boehner and Obama later that day at the White House and was surprised to see that Cantor wasn't aware of it. The meeting was being kept a secret, but Biden figured that Cantor was in the know.

The previous weekend, Obama and Boehner had played a high-profile round of golf at the course at Andrews Air Force Base. That outing had led them to schedule this private Oval

Office tête-à-tête to explore what sort of debt-ceiling deal might be possible.

Obama had not told top White House aides because he did not want to disrupt the Biden talks. And Boehner had his own reason for the secrecy. House Republicans, especially those eighty-seven die-hard freshmen, were still smarting from the CR fight, and many never stopped worrying that Boehner might cut another deal with the big-government socialist in the White House that would be a sellout of their cherished smother-the-government principles.

But Boehner not telling Cantor? Biden found that strange.

"The whole scene on the Republican side was weird," a senior administration official later said. "We really had no idea at times who we were dealing with."

THE NEXT DAY CANTOR CALLED THE VICE PRESIDENT'S OFFICE and left a message that he was pulling out.

This stunned Biden and the Democrats. It also surprised Boehner. Cantor had not told the speaker he was withdrawing.

Throughout the negotiations, Cantor had regularly appeared at House GOP caucus meetings to report progress. "He'd come in with decks showing all the cuts he was squeezing out," recalled a senior House Republican. "But when the talks were getting real and involved revenues, he didn't want to own it."

For Cantor, the hush-hush Obama-Boehner meeting was cause to worry that the speaker was undercutting his position in the Biden talks. Moreover, this was a good moment for Cantor to skedaddle, before he got stuck in the tar of revenue increases. He would let Boehner get in trouble for any compromises on that front.

THE BIDEN NEGOTIATIONS HAD YIELDED A LIST OF CUTS IN DIS-
cretionary defense and nondefense spending to the tune of about
$1.1 trillion, reductions in non-health mandatory programs (farm
subsidies, college aid, federal worker pensions, and more), and savings in Medicare and Medicaid. The negotiators had talked about
raising the Medicare eligibility age, means-testing the program,
and wringing money out of Medicaid by shifting costs to hospitals,
states, and providers. They had considered ending mail delivery on
Saturdays. But few details had been nailed down for much of this.

"It was like a menu," a White House official later said. "If we
do this, then we can do that or that."

The negotiators had delved into the complicated plumbing of
the federal budget and developed a substantial set of cuts totaling,
depending on how you counted it, about $1.6 trillion—which
fell short of the $2.4 trillion needed to meet Boehner's demand.
But it was all conditional. And the Republicans wouldn't fill the
hole with new revenues. For all their toil, the Biden group ended
up stuck.

The Republicans, naturally, responded to the Cantor walkout by . . . blaming the president.

"For weeks, lawmakers have worked around the clock to
hammer out a plan that would help us to avert a crisis we all know
is coming," McConnell bellowed. "So it's worth asking, where in
the world has President Obama been for the last month?"

This was misleading hyperventilation. Obama had sent
Biden and his senior economic advisers to bang out a budget
accord. The president's involvement, or lack thereof, was not
the issue. The dispute was over revenues. At the White House,
Carney said, "These talks are in abeyance."

BUT THERE WAS A NEW TRACK. THE SECRET OVAL OFFICE MEET-
ing between Obama and Boehner prompted a series of covert

negotiations. The president and the speaker had delicately discussed the possibility of a big deal, in which Obama would sign off on significant entitlement reductions while Boehner would yield on significant revenues. And they agreed that a small number of their respective aides would conduct follow-up conversations.

Late that week, Jack Lew, who was in New York, flew back to Washington—not telling his aides why—so he and Rob Nabors could secretly meet with Barry Jackson. In the White House, other senior aides did not know what Lew and Nabors were up to.

The initial conversations indicated that Boehner might consider a revenue boost of $800 billion achieved through tax reform (the elimination of loopholes and deductions), not the raising of rates—though there was the bedeviling issue of what baseline to employ (a wonky issue that made all the difference in the world). Given that Congress could move quickly on cuts but would need time to produce tax reform legislation—and there was no surfeit of time at the moment—both sides would have to agree on an enforcement mechanism, a so-called trigger.

If tax reform did not yield the promised $800 billion revenue hike, this backup measure would generate that amount. The trigger had to be unsavory enough to compel the Republicans to stay true to the tax side of the deal.

Boehner indicated he would consider a trigger that decoupled the Bush tax cuts for the rich from those for the middle class. This would allow Obama and the Democrats to extend only the tax breaks for the middle class and gain close to $800 billion in revenues. (If tax reform did happen, the Bush tax cuts would disappear, with new tax rates and rules established for all taxpayers and corporations.)

In return, Boehner was requesting a long list of entitlement changes: increasing the eligibility age for Medicare, rejiggering the cost of living adjustment (known as a COLA) for Social

Security recipients to slow the increase in benefits, direct cuts in Medicare and Social Security benefits, higher copayments for Medicare, and more.

After conversations with their GOP counterparts, Lew and Nabors met with Geithner, Sperling, Plouffe, and Reed in Daley's office and briefed them on their back-and-forth with Boehner's office. The question was whether the president should seriously engage with Boehner.

The $800 billion was far better than the zero revenues the Republicans had offered during the Biden talks. But was that enough—and would it seem enough to Obama's party? After all, if Obama merely blocked the extension of the Bush tax cuts for the wealthy (as he had vowed to do), he could produce roughly the same amount—and in the deficit-reduction framework he had unveiled in April, Obama had included hundreds of billions of dollars in additional revenues from taxes on top earners. Could he settle for less than that?

Then again, who knew what would happen during the next skirmish over the Bush tax cuts? A bloc of Senate Democrats could side with the Republicans to extend all the Bush tax cuts—or back a compromise ending the Bush tax cuts only for the *very* wealthy (which would not yield as much revenue). Now the White House had a chance to lock in $800 billion.

Boehner had a similar problem on his end. Compared with the Ryan budget, this deal would be pretty good for the president.

Obama's top economic advisers agreed they should seize this opportunity. They knew they could not give Boehner everything he requested on entitlements. They calculated they would have to concede one or two iconic issues, and they opted for raising the Medicare eligibility age and adjusting the Social Security COLA. To get an overarching agreement, a senior administration official later said, "these were the poisons we had to pick."

The president's aides figured that Obama's health care over-haul would help ameliorate the negative consequences of lifting the age for Medicare—and these changes, they believed, would not undermine the basic structures of these two critical programs.

The president was particularly torn over the Social Security concession. Lew told him there was a way to adjust the COLA to protect older seniors and even yield an increase in benefits for the poorest recipients. Lew was confident he could fiddle with the numbers to make it acceptable on a policy level—though Obama and his aides realized it would still be an explosive politi-cal decision.

Obama yearned to move past the never-ending budget-and-deficit scuffles and show investors and the markets—whoever they were—that the US government could address its long-run fiscal challenges. And deficit reduction was also necessary for cre-ating the political space he needed to pursue his future-oriented investments.

Obama and his advisers realized that downsizing government programs could possibly create a drag on the economy. But they were looking for a game changer to lift the economy out of the doldrums. A major deal might do this by removing uncertainty and encouraging confidence among businesses and consumers. Confidence, though, was a tricky business. It could be derided as the last-gasp defense of any policy. Yet at this point, Obama and his aides were searching for something that could flip the economic script.

Were Obama to broker such an ambitious bipartisan accord—overcoming the fierce opposition that would come from both sides—he could claim to be a postpartisan and pragmatic leader who put results ahead of ideology and party orthodoxy. He would be Washington's adult in chief, who pulled together the bickerers of the capital for the mother of all deals. It would be a monumental act of consensus crafting. Independent voters,

David Axelrod and others assumed, would just love it—perhaps enough to reelect Obama during a period of historically high unemployment.

The president and his team bore the responsibility of preventing a default, and Boehner had established a high bar: $2.4 trillion. Clearing that height would require unpalatable actions. Obama thought that if he and Congress were going that far, they might as well suck it up, suffer the extra political grief, and slog all the way to a grand bargain.

He instructed his team to proceed with the secret Boehner talks. He was hoping he and the speaker—two very different types of politicians who hailed from two very different political traditions—could fashion a historic compromise that would resolve several long-running policy battles of the nation's capital, restore the country to a sound fiscal footing, and demonstrate that Washington could function.

He might have been hoping for too much.

OBAMA WAS ALWAYS JUGGLING. THE DAY HE MET SECRETLY WITH Boehner, he was overseeing the military action in Libya, and he announced a crucial decision on the Afghanistan war.

At 8:00 that night, June 22, Obama addressed the nation and declared that in the next month he would begin withdrawing troops from what had become the longest war in US history, costing the lives of over seventeen hundred US service members and more than $1 trillion. Ten thousand troops would be removed by the end of the year, and another twenty-three thousand by the following summer.

The withdrawal would continue "at a steady pace" after that. Obama was de-surging to meet a target date of 2014 for handing over full security responsibility to the Afghan government and its military and police forces.

"The tide of war is receding," Obama said, adding, "America, it is time to focus on nation-building here at home."

The president, over the reservations of his top military advisers, was delivering on a promise he had declared eighteen months earlier.

IN DECEMBER 2009, OBAMA, AFTER A MONTHS-LONG, INTENSIVE review of US policy on Afghanistan and Pakistan, had ordered an influx of 30,000 troops into Afghanistan. This would raise the number of US troops there to about 100,000, almost triple the 34,000 troops on the ground when Obama inherited the war. These troops, he maintained, would target the Taliban insurgency, secure key population centers, and contribute to the training of Afghan security forces—all to establish better conditions for the United States to increase the pace of transferring security operations to the Afghans.

The US interest, Obama explained then, was in disrupting al-Qaeda and in ensuring that the Taliban was in no position to overthrow the Afghan government and once again provide al-Qaeda or other anti-American jihadists a safe haven.

In announcing this surge of troops, which would cost about $30 billion, Obama pledged that in eighteen months he would start to bring these troops home. The US commitment in Afghanistan, he said, "would not be open-ended."

As part of the strategic review, General Stanley McChrystal, then the commander of US forces in Afghanistan, had requested forty thousand extra troops (with no foreseeable end to their deployment) to be part of a broad counterinsurgency mission. McChrystal had warned in a report to the Pentagon that the United States risked failure without this boost in troops, and that report had been leaked to Bob Woodward at the *Washington*

Post—a move some White House officials regarded as an attempt to bully Obama into okaying McChrystal's request.

The ten-year cost for McChrystal's request was possibly $1 trillion, and for Obama this eye-popping price tag and the prospect of a decade of further involvement in Afghanistan was too much—especially for a war that might be beyond traditional victory. Obama and most of his top national security aides assumed that there could be no total defeat of the Taliban, an indigenous force integrated into the fabric of Afghan society. The endgame would have to be some sort of political resolution or a situation with the Taliban existent but not strong enough to threaten seriously the central government.

But Obama had come to believe it made sense to throw extra forces at the Taliban temporarily to knock it back and afford the Afghans the opportunity to beef up their own capacity and take over more of the fight. He altered the plans McChrystal and the Pentagon had presented him, opting for a smaller number of additional soldiers and instructing the military to deploy these new troops at a faster rate so they could be withdrawn sooner. Obama had essentially drawn up his own blueprint: there would be more troops to pave the way quickly to fewer troops. If this worked.

It was a decision that would satisfy neither the hawks nor the skeptics. This was not a battle plan to achieve complete triumph, and John McCain and other champions of expanded military engagement in Afghanistan derided the notion of setting a withdrawal date as defeatist and strategically counterproductive.

Critics of the war saw Obama upping the investment in an enterprise that was beyond salvaging, especially considering the rampant corruption of the government and the ineptitude of the Afghan security forces.

As Obama discussed his decision with aides and military commanders, he made sure they all understood he was serious about the July 2011 withdrawal date. That's when the surge would reverse. He would not consider additional troops. Yet a week after Obama announced the additional troops deployment, McChrystal testified on Capitol Hill: "I don't view July 2011 as a deadline. At that time, we'll evaluate the time and scope of a possible drawdown."

A *possible* drawdown. It looked as if the military had not taken the president seriously.

About this time, Senator Carl Levin, the chairman of the Senate Armed Services Committee, saw the president and warned him: They're trying to "make mush" of your date. They're trying to turn your deadline into a goal.

"It's an order, not a goal," Obama replied. The withdrawal would happen.

In the following year and a half, whenever Levin saw the president, he always pressed Obama to build up the Afghan military and adhere to the July 2011 date. Obama constantly assured Levin the deadline was firm.

WITH THE SURGE UNDER WAY, THE WAR CHUGGED ON WITHOUT much public debate. It was rarely in the news. It never became an issue in the 2010 midterm elections. During this time, the dicey relationship with the erratic President Hamid Karzai, who had been reelected in a contest tainted by fraud, continued to be a strategic nuisance for US policymakers. Widespread corruption remained an obstacle. Afghan security forces grew and received greater training—but questions regarding their effectiveness persisted.

Yet US and NATO military forces were able to push the Taliban from several critical areas, and the US counterterrorism efforts—including the CIA's not-that-secret drone program targeting al-Qaeda and related outfits in the tribal regions of Pakistan—severely set back bin Laden's operations.

Come spring 2011, with the due date for the initiation of the drawdown looming, Obama had to decide how many troops he would remove and at what pace. In December 2009, he had said that the number would depend on the conditions at the time. Now it was time to make the call.

In a Washington consumed with deficit reduction and spending cuts, congressional support for the war was ebbing. In mid-May, a House measure requiring that Obama establish a firm timetable for withdrawing US forces from Afghanistan lost on a close 215–204 vote, with dozens of Republicans joining progressive Democrats to oppose the war.

In the following weeks, signs of unease continued. Levin was calling for withdrawing fifteen thousand troops by the end of the year. Twenty-seven senators, including two Tea Party Republicans, urged Obama to proceed with "sizable and sustained" reductions.

Senator Richard Lugar, who was not part of this group, observed, "It is exceedingly difficult to conclude that our vast expenditures in Afghanistan represent a rational allocation of our military and financial assets."

As he considered the size of the withdrawal, Obama conducted a narrow set of deliberations. After what happened in 2009—a long review process that produced leaks and an apparent effort on the part of the military to box him in—the president was not interested in another comprehensive interagency evaluation.

He held individual meetings with his defense secretary and his secretary of state, conferred with his national security staffers, and chaired three full NSC meetings in the two weeks before announcing his decision.

In all these discussions, Obama was unambiguous: the reductions of troops would be meaningful, not perfunctory; the transition to Afghan control would be significant, not for show.

Obama was leaning toward removing all the surge troops by March 2012—within nine months. Robert Gates, Hillary Clinton, and General David Petraeus, who now commanded US forces in Afghanistan (and whom Obama had recently nominated to become CIA chief), preferred slower and smaller reductions. Gates, who would soon be leaving the Pentagon, was advocating an eighteen- to twenty-four-month time line for bringing home the surge troops. And he was hinting as much in public. Petraeus urged Obama to remove only five thousand in 2011 and the same amount in 2012.

Military officials contended that a larger withdrawal would not allow them to protect the gains they had achieved in key provinces. (In one leak, military sources told the *Wall Street Journal* that removing more than ten thousand troops would undercut the progress they had achieved.)

Obama, Levin later noted, "was under great pressure to draw down no more than 5,000 troops in the first year and another 5,000 to 10,000 by the end of 2012."

Obama stuck with the larger number: all thirty thousand surge troops (plus a few thousand support troops). But he was flexible on the speed of the drawdown. Petraeus asked that these troops not be entirely withdrawn until the end of 2012—nine months beyond the March 2012 deadline Obama had in mind.

Gates proposed September 2012 as a compromise. This would keep the troops in Afghanistan through most of a second fighting season. Obama agreed.

Resisting the calls from Gates, Mullen, and Petraeus for a less consequential withdrawal, the president—who certainly was in a stronger position to deny the Pentagon after the bin Laden raid—stayed true to his pledge to add troops only on a temporary basis.

OBAMA'S SPEECH ON JUNE 22, 2011, UNVEILING THE WITHDRAWAL schedule was meant to be an unequivocal signal that the war was heading toward a conclusion. Still, the president was defending a not-yet-over conflict that was unpopular—and he was promoting a complicated message that might come across as contradictory: this war was essential for the security of the United States, but the US commitment to it was limited.

Regardless of the number of troops Obama was ordering home, the war remained a hard sell. Some of the progress that the military claimed was debatable. The capabilities of Karzai's corrupt government and Afghan security forces were still uncertain. Then there was the question of the rationale for the whole endeavor.

During a background briefing hours before Obama's speech, a senior administration official told reporters that "we haven't seen a terrorist threat emanating from Afghanistan for the past seven or eight years" and that only fifty to seventy-five "al-Qaeda types" were operating within Afghanistan. These fighters, he explained, were merely members of "tactical fighting units inside of Afghanistan." There was "no indication at all" of any al-Qaeda efforts within Afghanistan to use it "as a launching pad to carry out attacks outside of Afghan borders."

At the same time, this official said, the United States' efforts had damaged "al-Qaeda's core capabilities significantly." Washington's secret warriors had killed senior leaders of al-Qaeda in addition to bin Laden. Waziristan in Pakistan was no longer a safe haven for al-Qaeda. Drone attacks and special forces assaults targeting terrorists were continuing.

This official summed it up: "We don't see a transnational threat coming out of Afghanistan in terms of the terrorist threat."

This sounded like a declaration of victory and raised the question: If the overall goal of the Afghanistan mission was to disrupt, dismantle, and defeat al-Qaeda and protect the United States from attacks that could be devised and launched from Afghanistan, was it still necessary to wage a full-scale war to achieve that end?

White House aides noted that the United States had to guarantee Afghanistan didn't reemerge as a safe haven for a down-but-not-out al-Qaeda, and this meant ensuring, as one staffer put it, "a degree of stability in Afghanistan" and a government that "could stand on its own two feet and not be overrun by the Taliban." But could these conditions be met in the next two years?

OBAMA WAS ONCE MORE POSITIONING HIMSELF BETWEEN THE ardent champions of the war and its exasperated opponents.

McCain, predictably, led the charge from the right: "I am concerned that the withdrawal plan that President Obama announced . . . poses an unnecessary risk to the hard-won gains that our troops have made thus far in Afghanistan."

Critics of the war pointed out that even after this withdrawal, Obama would still have more troops in Afghanistan than he had at the start of his presidency.

"It has been the hope of many in Congress and across the country," Pelosi remarked, "that the full drawdown of US forces would happen sooner than the president laid out." The Center for American Progress noted that "many important questions about Afghanistan, including our core objectives, future costs, how military operations will support the political and military transition between now and 2014, and our relations with Pakistan, remain unanswered." It called for pulling out sixty thousand troops by the end of 2012.

But Obama's decision didn't lead to much of a firestorm. The disappointed hawks seemed uneager to stir too much of a ruckus, perhaps because polls continued to show the costly war was unpopular.

Obama, his aides believed, was in a sound place; he was ending the wars in Iraq and Afghanistan—even if too slowly for some—without having to worry that large portions of the voting public would perceive him as weak on national security and not interested in victory. (Later in the year, he would withdraw all US troops from Iraq, in accordance with an agreement negotiated by the Bush administration and the Iraqi government.) The risky bin Laden operation had proved that he was damn serious about killing America's enemies.

"We could shrug off the far Right," a senior administration official said. "And to the far Left we could say we share the goal of ending the war. This was a good spot."

Obama's plan for Afghanistan was one more reflection of his pragmatism. He had factored resource allocation and political will into the equation. He had focused more on finding a measured path to a satisfactory outcome than on achieving an ideal goal.

His counterterrorism strategy was not aimed at grand geostrategic initiatives. It was narrow and concentrated on individual terrorists—killing them with drones and other means (which in some instances led horribly to civilian casualties). And his political desire was to conclude the Afghanistan war without being seen as opposed to the war.

The overarching message was that the 9/11 era was over. A decade later, after killing Osama bin Laden, Obama was winding down the wars that had been launched in response to bin Laden's attack on the United States.

Obama and his national security aides hoped to recalibrate US foreign policy to place more emphasis on the Pacific region—even as they would continue to contend with the Arab Spring, Iran and its nuclear program, the anemic Middle East peace process, and the ongoing effort to crush al-Qaeda and its allies in the Af-Pak region and beyond.

Yet determining how many troops to withdraw over the next fourteen months would not remove Afghanistan from Obama's to-do list. The demanding tasks of collaborating with Karzai and handling the assorted complexities of Pakistan would not cease. Afghan government corruption and military ineptitude would not quickly fade as threats to stability and progress. (In early 2012, the Pentagon would announce its intention to end combat operations in Afghanistan in mid-2013, a year ahead of schedule.)

Obama had placed US engagement in Afghanistan on a gradual downward slope. But the war was far from done.

Chapter Twelve

THE BIG CROSS

AS OBAMA PLOTTED OUT THE NEXT ROUND OF DEFICIT TALKS with John Boehner, he had settled on a saying borrowed from Joe Biden—"Don't die on a small cross"—to encapsulate his thinking.

He knew he had to reach a deficit-reduction deal to prevent a default (and economic peril). He believed that addressing the nation's long-term fiscal challenges would enhance economic prospects over the long run. And he realized that any negotiated agreement might help him with independent voters, but it would also come with a boatload of political pain.

After watching all the trouble Biden had gone through to reach just a medium-sized deal, Obama had concluded he might as well shoot for the moon and aim for an accord that could resolve the deficit issue for years to come.

Biden agreed with the president: go for the big cross. There would be enough political gain to buffer that pain.

"In the end," the vice president said more than once at White House meetings, "I don't think you die on it."

ON JUNE 29, OBAMA HELD A PRESS CONFERENCE TO TOUT HIS recent and modest steps to buoy the employment picture—a government-wide review of regulations and a community college job-training initiative. But the question of the day was whether he could pull off a debt-ceiling deal.

Obama reiterated the basics from his anti-Ryan speech. Then he teed off on Congress for taking recesses while the debt-ceiling issue was unresolved: "Malia and Sasha generally finish their homework a day ahead of time. Malia is 13, Sasha is 10. . . . They don't wait until the night before. . . . Congress can do the same thing."

Obama was turning all of Congress into a punching bag, not just the Republicans—a point not well taken by his Democratic allies on the Hill. He was lumping the Rs and Ds together in something of a Capitol kindergarten.

This would be his tune for the coming weeks. He would often blame all of Congress for Washington's woes. (This was not an unwise strategy; in a recent poll, just 17 percent of Americans approved of Congress's job.)

But Capitol Hill Democrats were becoming increasingly impatient with Obama's restraint regarding Republicans, complaining that he was not aggressively framing the debt-ceiling controversy and calling out the Republicans for playing Russian roulette with the economy. After all, if Obama would not depict the GOPers as irresponsible ideologues driving the nation to the brink of financial crisis and possibly another recession, who would? Nancy Pelosi? Harry Reid? Tim Geithner? No one else had the president's wattage.

The day after the press conference, Reid announced he was canceling the Senate recess scheduled for the coming week. The House was already due to return to work then.

THAT HOLIDAY WEEKEND, OBAMA AND BOEHNER HELD AN-other secret meeting at the White House. Sitting on the patio near the Oval Office, they reviewed the contours of the grand bargain. Staff from each side followed up with additional conversations. And this was still all on the QT. Boehner didn't want his Republican troops to know he was in cahoots with the president.

In public, Boehner showed no signs he was negotiating with the president and discussing concessions on revenues. In fact, the speaker even issued a statement denouncing Obama for requesting tax-reform-related revenues in a debt-ceiling bill—precisely what his staff and White House aides were working on.

Having begun covering the basics of an ambitious agreement with Boehner, Obama asked the eight top congressional leaders—Harry Reid, John Boehner, Mitch McConnell, Nancy Pelosi, Dick Durbin, Eric Cantor, Jon Kyl, and Steny Hoyer—to meet at the White House on July 7 for talks that he hoped would get this big thing done.

The night before the congressional leaders were to arrive, the grand bargain was unveiled—in a *Washington Post* article that leaked the preliminary details of Obama and Boehner's potential deal. The paper reported Obama would be pitching the congressional leaders a $4 trillion agreement that would include "major changes" in Medicare and Social Security in return for "fresh tax revenue."

This article set off alarms throughout Washington—especially among Democrats, who worried that Obama was placing his party's treasured programs on the chopping block to reach an accommodation with the intractable Republicans.

AT 11:00 A.M., THE EIGHT CONGRESSIONAL LEADERS FILED INTO the Cabinet Room, which is next to the Oval Office and over-looks the Rose Garden, and sat down at the long oval mahogany table. Obama led the meeting, with Joe Biden, Tim Geithner, Jack Lew, Gene Sperling, and Bill Daley joining him.

The president said the debt matter must have a bipartisan resolution, and he would not accept any short-term debt-limit extensions. He was not willing to go through this rigmarole again. He would veto any measure that raised the debt ceiling for only a few months.

Geithner emphasized that the August 2 deadline was drop-dead real. Treasury officials had already pushed the date back, but the creative bookkeeping was exhausted. A hundred billion dollars in debt would come due every week after that. And, the Treasury secretary added, the Fourteenth Amendment was no "magic bullet."

Some Democrats had urged Obama to invoke the Fourteenth Amendment—which states that the "validity of the public debt of the United States . . . shall not be questioned"—to skirt the debt-ceiling limit and instruct the Treasury Department to borrow money on its own to cover the government's bills. Treasury officials had investigated the option and concluded it just might be possible.

But this option was open to constitutional interpretation. Article I of the Constitution plainly states that Congress possesses the power to borrow money. And legal arguments aside, there were practical concerns. If Obama tried this, he would undoubtedly face an impeachment effort in the House. More important, any attempted borrowing would occur under such a large cloud of legal ambiguity that lenders would not have confidence in the bonds being issued. Interest rates could soar. It would be tumultuous. The president and his aides never considered this seriously.

Geithner's point was simple: without legislation, there was no easy way out.

Next, Obama, adopting his professorial manner (which often agitated Republican leaders), stated that they had three approaches from which to choose. He ran down the list, outlining the pros and cons of each.

The small deal included a package of discretionary cuts and reductions in mandatory programs, with no revenues and no whacks at Medicare or Medicaid and a debt-limit boost that would run through 2012. This package would likely not meet Boehner's one-for-one standard.

The second was the middle course that the Biden group had explored. This would include some revenues and some entitlement cuts.

Then there was the grand deal: cuts in discretionary and mandatory spending, significant reductions in entitlements without any Ryanesque changes in the nature of these programs, more than a trillion dollars in revenues, and a commitment to tax reform.

The president favored the supersized deal. They had to get this debt matter over and done with, he maintained, so they could all turn their attention to another fight: what to do about the economy and creating jobs now.

"No sense dying on a small cross," Biden said.

It was then time for the others to state where they stood.

"I didn't run for speaker just to have a fancy title," Boehner said. He was for a grand bargain.

Reid and Pelosi indicated that they were open to the big cross—but with big ifs.

"It has to be balanced," Reid said, "between spending and revenues—in terms of timing, specificity, and dollars."

In other words, it couldn't be cuts now but revenues later.

Pelosi also had a line in the sand: no cuts aimed at Medicare beneficiaries. But savings on the provider side could be acceptable.

The House Democratic leader realized that any compromise that included Medicare reductions (even if only affecting providers) would undercut the Democrats' ability to attack Republicans for voting to end the Medicare guarantee. But she and her political advisers believed the Democrats could still make the case that the GOPers had attempted to dismantle Medicare, while the Democrats had supported cost savings that would preserve the program for years to come. (Plouffe thought so as well.)

Cantor and Kyl each expressed opposition to the grand bargain. Cantor insisted it had to be midsized. Both said that a large agreement would never pass the House. Kyl also noted he was opposed to serious cuts for the Pentagon—which would have to be part of a big-cross deal.

McConnell, perhaps the cagiest of the leaders, was supportive of Boehner's ambition to go big. But he spoke the least—which was not unusual—noting that a large deal would be fine with him, though unlikely to survive in Congress. The Democrats couldn't help noticing each number two on the GOP side was out of sync with his leader's stated desire for a grand bargain.

It was three months since the lawmakers had met with the president before his fiscal policy speech, a senior administration official later pointed out, and the president and congressional leaders were still not in agreement on the size of the cross.

THIS CABINET ROOM MEETING SHOWED THAT THE WHITE House was once more not dealing with a unified Republican Party. Was Boehner really in a position to negotiate a grand bargain—which would require holding the Tea Partiers at bay—if his second in command, who was closer in spirit and style to this far Right band, was not aboard, and if his comrades in the Senate were unenthused by the prospect?

"Perhaps that was the moment we should have known better," a White House aide later said.

Democrats were fretting about Medicare and Social Security being in play. Back on Capitol Hill, Pelosi declared, "Any discussion of Medicare or Social Security should be on its own table. We are not going to balance the budget on the backs of American seniors, women, or people with disabilities."

Capitol Hill Democrats were suspicious of Obama. Was he hoping to win back those unreliable independent voters by sacrificing entitlement programs his party had long championed and defended? How far was he willing to go?

At a closed-door meeting of House Democrats, Representative Steve Israel, who now chaired the DCCC, declared, "We will defend Medicare and go our own way from the White House, if we have to."

Obama was risking a split in his party. If he got close on the grand bargain, he could ignite a civil war among Democrats.

AFTER THE JULY 3 MEETING BETWEEN OBAMA AND BOEHNER, Lew and Nabors had begun haggling with Jackson and Brett Loper, Boehner's policy director, about the details of a big deal. They concentrated on two critical issues: the trigger and the basic shape of the tax reform (which would supposedly generate $800 billion in revenues).

Boehner's office sent the White House a proposed tax reform plan, which Obama, Geithner, Sperling, Reed, and others concluded was not sufficiently progressive. The White House sent back a response and expected a quick counteroffer. They seemed to be on their way.

But as Boehner's aides were negotiating with the White House, members of the House Republican caucus grew uneasy about the prospect of Boehner coauthoring a grand compromise with Obama. They realized that meant yielding on taxes. The

Tea Party Republicans had come to Washington to smash the status quo, not cut deals.

"They were giving Boehner a helluva time," a senior House Republican staffer recalled. "They weren't interested in deal making. They didn't want to recognize this was divided government. They were willing—eager—to see what would happen if they refused to compromise."

Several of Boehner's closest allies in the caucus came to the speaker's office. They knew Boehner was ready to compromise with Obama on revenues and contemplating Obama's proposal to decouple the upper-income Bush tax cuts as the trigger.

"We told him, 'You're too far over the tips of your skis,'" one of these lawmakers later said.

Boehner, they feared, was pursuing an agreement that his members couldn't stomach, and that meant his speakership was in jeopardy. Cantor and his allies, they told him, were whispering that Boehner had gone RINO—that much-dreaded insult in conservative circles: Republican In Name Only.

"Cantor," one of these members subsequently maintained, "is a palace-intrigue guy. He was just waiting for a bus to hit Boehner so he could say, 'Oh, I wasn't planning on being speaker, but if I have to be. . . .'"

Boehner's allies told the speaker he was at risk of losing not 40 to 60 Tea Party members of the caucus, but 150 members. If that occurred, his speakership would be kaput. Cantor would lead—or benefit from—a revolt. There was not much percentage in reasoning with the Tea Party wing and their allies in the leadership, Boehner's friends explained to the speaker.

"The young guns really believe that if everything blows up, there will be a Republican president, a Republican House, and a Republican Senate," one of these GOP lawmakers subsequently said, "and they'll be able to do whatever they want."

ON JULY 9, A SATURDAY, OBAMA'S SENIOR AIDES—LEW, SPER-
ling, Nabors, Reed, and others—were waiting to hear back from
Boehner's office. In the afternoon, the White House put out a
call to these top officials: Daley wants you in his office as soon as
possible. As Sperling headed toward 1600 Pennsylvania Avenue,
he was excited. He thought this meant a deal was close, and he
started pondering what final details had to be confirmed in the
next twenty-four hours.

When Sperling and others arrived at Daley's office, Nabors
handed out a statement Boehner had just released: "Despite
good-faith efforts to find common ground, the White House
will not pursue a bigger debt reduction agreement without tax
hikes. I believe the best approach may be to focus on producing
a smaller measure, based on the cuts identified in the Biden-led
negotiations, that still meets our call for spending reforms and
cuts greater than the amount of any debt limit increase."

Boehner had called Obama at Camp David to tell him he
was no longer pursuing the grand bargain.

Obama and his aides were stunned and puzzled. Who ends
negotiations this way? Senior staff had been exchanging paper
and proposals. They were working it out. Yet Boehner had
jumped ship without notice. It was bizarre.

Obama's aides did not know that Boehner had been told that
Cantor and other Republicans were waiting to pounce on him.

But there was a clue as to what was transpiring on the other
side. That morning, the conservative editorial page of the *Wall
Street Journal* had warned Boehner against cutting a deal with
Obama. The editorial focused on the possible trigger of decoupling the Bush tax cuts and suggested that Obama had outmaneuvered Boehner. Why, the editorialists asked, would Boehner

agree to a possible tax hike before tax reform was done? Wouldn't that provide Democrats the incentive to stall tax reform and grab the tax-cut win that had so far eluded them?

The article, which hinted that Boehner's speakership would be on the line if he green-lit such a deal, was laced with so many details from the negotiations that someone within the House Republican leadership must have orchestrated it. It was a sign that Republicans on the inside were gunning for Boehner.

"There was a feeling that the speaker was going further than his caucus would," a senior House Republican staffer subsequently explained. "We heard rumors a tax increase was involved. I know my boss made clear to him the conference would not support that."

White House aides sensed that Boehner had hoped to sneak a deal past the Republican caucus. At one point, a Boehner staffer had said to White House aides, "We can do better on revenues as long as you don't describe what we're doing as revenues." That told the Obama advisers a lot.

THE NEXT EVENING, OBAMA HAD ALL THE CONGRESSIONAL leaders back to the White House. He told them he was willing to go far to get the big-cross deal, noting he'd accept a gradual rise in the eligibility age for Medicare.

"If not now, when?" he asked.

There's not enough time, Boehner said—reversing his position from three days prior.

Reid showed his irritation. The Republicans, he complained, had backed out of one deal after another. He ticked off the instances. Senate Republicans who had cosponsored a bill to set up a congressional deficit commission voted against it after Obama endorsed it. It was a Republican who had run away from the

Gang of Six talks. Cantor had left the Biden negotiations. And now Boehner had turned tail on the big deal.

Cantor noted that the grand bargain breached his and Boehner's obligation to their caucus. In other words, they couldn't sell any package that contained revenue increases.

McConnell barely said a word, but Kyl suggested that they focus on the Biden group savings that did not make "fundamental changes" in Medicare. He was trying to claim the cuts the Biden gang had discussed, sidestepping the fact that they were only provisional without accompanying revenues.

Obama reminded the Republicans it had been their idea to deal with deficit reduction at the same time as the debt limit. Now, he said, they were running scared.

"It was a completely unproductive meeting," a Democratic staffer recalled.

WASHINGTON HAD REACHED A CURIOUS POINT. A DEMOCRATIC president was pushing for more deficit reduction than the Republican leaders of Congress were. And Obama was beginning to win the message war: he was willing to compromise to reach a balanced and big deal, but the Republicans were too wed to tax cuts for the well-to-do to make anything happen. An NBC/ *Wall Street Journal* poll found that Obama's approach to deficit reduction was favored over the Republicans', 58 to 36 percent. In a Quinnipiac poll, 67 percent of voters agreed that legislation to lift the debt ceiling should include tax hikes for corporations and the wealthy, not just spending cuts. Only a quarter opposed such a move.

At a press conference, Obama tried to nudge Boehner back toward the grand bargain. He said he was open to "meaningful changes to" Medicare, Medicaid, and Social Security, but he expected Boehner and the Republicans to reciprocate with revenues. He noted that "nobody has talked about increasing taxes"—just eliminating tax breaks and closing loopholes. He repeatedly praised Boehner as a sincere fellow who had been acting in good faith. Obama said the speaker's caucus was the problem.

The president also pleaded with progressives to recognize that reforms were necessary for these cherished entitlement programs to survive. Medicare, he noted, was projected to run out of money eventually: "If you're a progressive who cares about the integrity of Social Security and Medicare and Medicaid . . . we have an obligation . . . to make it sustainable over the long term." And, he added, if the deficits weren't tackled, spending for investments that liberals supported—Head Start, student loans, medical research, infrastructure, and all the rest—would be threatened.

This was a case that tended not to fly with Obama's liberal allies. When the economy was lousy, when millions were out of work, when millionaires and corporations were doing just fine, when Wall Street had recovered and the banks were back on top, why was it a top priority to slice these mainstay programs for middle- and low-income families? (And wasn't the real long-run problem the cost of health care, not the cost of Medicare? Squeezing Medicare would just transfer those out-of-control costs to vulnerable beneficiaries.)

Many Democrats and progressive activists had trouble viewing this as anything but capitulation to the Tea Party extremists or a political move designed to impress independents who had fallen out of love with the president. But the Republicans were holding the debt ceiling hostage, and Obama was offering his side the opportunity to establish a fiscal regime that would secure these programs for years. The president was also suggesting that it's better for us to do it than to leave it to *them*—particularly because this *them* could have total control of the government after the next election.

ON THE AFTERNOON OF JULY 11, OBAMA, BIDEN, AND THE CON-gressional leaders were back in the Cabinet Room. And Cantor tried to pull a fast one.

The meeting opened with the majority leader presenting charts representing cuts in non-health mandatory programs that had been discussed during the Biden talks.

You're missing a page, Obama quipped. The one with revenues.

Cantor moved on to Medicare and Medicaid reductions that he claimed the Biden group had accepted.

Reid called him on that. The Senate majority leader said he had spoken with Max Baucus and Daniel Inouye, and they said nothing had been settled on Medicare and Medicaid. The congressional leaders then began discussing possible cuts in these entitlement programs, but Reid stopped them: without significant revenues, there would be no health care changes. Durbin, Pelosi, and Hoyer backed him up.

Reid put another notion on the table: a $2.6 trillion package that would include cuts in discretionary spending and non-health-care mandatory programs and count the savings from winding down the wars in Afghanistan and Iraq. (Ryan's budget had included such a peace dividend.) Plus, there would be a joint congressional panel—a supercommittee—to produce an additional deficit-reduction package that would be put to a straight up-or-down vote in the House and Senate after the next election and inauguration.

Reid was proposing a midsized cross that could grow into a larger one.

Obama reiterated that he would not support any plan that

was not full of pain for both sides. Shared sacrifice, Obama said, was his bottom line.

At one point, Boehner challenged the president: "Look, entitlement cuts aren't easy for us to vote for either. Our guys aren't cheerleading about cutting entitlements."

"Your guys already voted for them!" Obama replied, referring to the Ryan budget.

Boehner shot back: "Excuse us for trying to lead."

Obama ended the meeting by remarking that he could not believe the Republicans would not ask wealthy folks to contribute anything.

McCONNELL, NO FAN OF THE GRAND BARGAIN OR THE WHITE House talks, had begun concocting a measure that would allow the president to lift the debt ceiling three times in the coming sixteen months unless Congress passed a veto-proof resolution of disapproval at each stage. Under this devilishly clever plan, Republicans could vote against extending the borrowing authority and there would be no default.

Obama and Democrats would assume responsibility for the debt-ceiling hike each time—and they would have to do so thrice before the next elections. It was political gamesmanship of the highest order.

ANOTHER MEETING OF OBAMA AND THE CONGRESSIONAL LEADers convened the next day. These sessions were taking on the stale air of uselessness. There was little mutual trust. The participants knew that anything they said would leak. The division on the Republican side was so pronounced the White House had a hard time even cooking up a strategy. White House aides were losing faith in the process.

And Boehner and McConnell were publicly blasting Obama, complaining he was not willing to cut deeply enough and was too enamored of tax hikes. "This debt limit increase is his problem," Boehner huffed, "and I think it's time for him to lead by putting his plan on the table—something that Congress can pass. Where is the president's plan?"

ON JULY 13, DALEY, GEITHNER, LEW, SPERLING, ROUSE, AND Reed gathered in the Oval Office to prep Obama for another session with the congressional leaders in the Cabinet Room.

"I want to make something clear," Obama told his aides. "I'm not going to accept a short-term extension of the debt."

He reiterated that he would not go through this exercise again in three, six, or twelve months.

His aides reacted with the sort of respectful but we-know-best condescension that advisers to the powerful sometimes muster: we understand why you feel that way, but when it comes down to it, we just may have to give in on this.

"I'm not doing it again," Obama said. "This is wrong."

The president felt that the Republicans' threatening default to get their way on budget issues distorted the separation of powers. It shouldn't be allowed. It was an embarrassment for the United States. It was not what the framers of the Constitution intended.

Obama was frequently disparaged for not adopting a firm stand. Here was his line. No more of this bullshit—at least not until after the next election.

OBAMA WAS TIRED OF TRYING TO CONVINCE THE REPUBLICANS to take a leap with him. At the meeting with congressional leaders later that day, he pointed out that he was personally attempting to work out the details of a big deal—far more than Reagan or George W. Bush would have done. He told the lawmakers the Republicans had shown him nothing that would push the conversation forward.

Obama, noting his pledge to veto any short-term extension of the debt ceiling, said, "This could bring my presidency down."

But, he added, "I will not yield on this." He would not allow the presidency to be held hostage again.

Throughout the meeting, Cantor and Obama tangled over the time frame of any debt-ceiling hike. Cantor was now advocating a smaller deal with a shorter extension that would lead to another face-off within a year.

The Democrats thought Cantor was especially prickly this day. He repeatedly interrupted the president. During the Biden talks, the vice president, Lew, and Sperling had developed a respect for Cantor. He had come across as serious, and there was no drama.

But when he was in the room with Obama, they observed, he became a different person. In these larger White House meetings—when leaks of what transpired would reach his fellow House Republicans (or Rush Limbaugh)—Cantor became more confrontational.

Toward the end of this session, Cantor once again raised the idea of a short-term extension.

"I'm not going to do it," Obama angrily said to Cantor. "We're not putting the country through this again. Don't call my bluff."

And he left the room.

Hours after the meeting, Cantor publicly claimed that Obama had stormed out after Cantor challenged him. He seemed to be boasting about getting under Obama's skin in an effort to score points with the Tea Party set.

THE MEETING THE NEXT DAY WAS NO BETTER. THE NEGOTIATORS were moving apart. McConnell had nothing invested in these talks. Reid and Pelosi backed the idea of a grand bargain, but were leery of Obama's entitlement concessions. (Reid had also started working with McConnell on a bill blending their two proposals—but it violated Obama's call for no short-term extensions.)

Lew, Sperling, and Geithner reviewed the work of the Biden talks and proposed closing loopholes for oil and gas companies, ending tax breaks for corporate jet owners, and removing subsidies for ethanol producers. The Republicans in the room rejected all of this. They were holding firm to their party's revenues-equals-tax-hikes orthodoxy. Cantor was mostly mum.

We've walked through all the components, Obama remarked. We have to figure out how to get to 218 votes in the House and 60 in the Senate. Go back to your caucuses and work that out.

"It's decision time," he said.

But it really wasn't.

DEFAULT ANXIETY WAS SPREADING. FEDERAL RESERVE CHAIRman Ben Bernanke warned that failure to raise the debt ceiling would amount to "a self-inflicted wound" that would cause "a very severe financial shock" to the global economy. China, the US government's largest foreign creditor, called on US policymakers to take action to protect the interests of investors.

OBAMA COULDN'T LET GO OF THE BIG DEAL.

"I've got reams of paper and printouts and spreadsheets on my desk," he said at a mid-July press conference, "and we know how we can create a package that solves the deficits and debt for a significant period of time." He was the guy who keeps chasing the girl who keeps saying no. But then romance was once again in the air.

That day, Boehner invited Geithner and Daley to his office. With Cantor by his side, Boehner presented a deficit-reduction plan that would include severe entitlement reforms and a tax code rewrite to generate $800 billion over ten years.

By now, White House aides were not sure Boehner could guarantee any offer he made. But Obama's senior advisers decided to play this one out. At least this time Cantor was by Boehner's side. The possibility of a grand bargain had risen from the dead.

Two days later—with the congressional leadership talks over—Daley, Geithner, and Lew were discussing the package with Boehner and Cantor at the White House. It would contain $1.2 trillion in discretionary cuts and procedures for finding trillions of savings via tax reform and entitlement reductions in the coming months. The White House would slice about $250 billion from Medicare (and gently raise the eligibility age) and trim Social Security benefits through the COLA change.

The ever-knotty matter of the trigger had yet to be resolved. The White House aides suggested sticking with decoupling the Bush high-income tax cuts. Boehner and Cantor nixed that. (That's what had gotten Boehner into trouble earlier.)

The two Republicans proposed a repeal of the individual mandate from Obama's health care plan. Obama's aides thought that was ridiculous. Or, Boehner and Cantor suggested, greater cuts in Medicare and Medicaid?

Boehner now was requesting more in entitlement cuts and offering less on the high-end taxes. This new deal was less attractive to the White House, but it was close enough. Obama's aides once more allowed themselves to believe a grand bargain was within reach.

"We had been going through this Shakespearean thing," a senior administration official recalled, "trying to negotiate with

THE BIG CROSS / 321

them, as Boehner and Cantor were trying to kill each other. Now they were back as a pair. More stingy. But we knew this was the last chance."

Staffers on both sides started working. In public, Obama and Boehner were tussling over the so-called Cut, Cap, and Balance bill, which the speaker was pushing. This measure would slash government spending by 25 percent and require passage of a constitutional amendment mandating a balanced budget before the debt ceiling could be increased. Looking to placate his Tea Partiers, Boehner had scheduled this red-meat legislation for a vote that week, knowing it stood no chance of success in the Senate. And Boehner's office was publicly denying he was once again negotiating with the president. As is often the case in Washington, there was the public game, and there was the inside game.

"The deal we were looking at was worse than what we'd accept in normal circumstances," a senior administration official recalled. "We were willing to take a suboptimal deal because we were governing, and with our economy on the line, someone had to be the adult. We couldn't allow our nation's first default or the significant chance we would feed the uncertainty in Europe to the point where we risked a recession at home."

Obama's aides wanted this deal badly.

BUT THEIR HOPES RAN INTO A WALL WHEN THE STILL-ALIVE GANG of Six in the Senate (Coburn had rejoined the group) unveiled a bipartisan deficit plan that would sop up about $4 trillion in red ink with the usual assortment of spending and entitlement cuts, but with about $1.5 trillion in additional revenues. A bipartisan group of forty senators attended a briefing on the plan and the vibe in the room was positive.

The problem for Obama was that this plan included far more revenues than the agreement he was brewing with Boehner and

Cantor. Their $800 billion in tax-reform-generated revenues would seem puny in comparison—and House and Senate Democrats would justifiably wonder why they should vote for legislation with significantly lower revenues than a package backed by several Republican senators.

ON THE EVENING OF JULY 20, OBAMA INVITED BOEHNER AND Cantor to the Oval Office. He told them that the Gang of Six had changed the playing field by demonstrating that Republicans could live with significant levels of revenues. Now he would have trouble selling their deal to congressional Democrats without more revenues. If the Republicans couldn't add anything, they would collectively have to scale back the entitlement reductions in the package.

This wasn't an ultimatum. The president was sharing a political reality and looking for a way to get past it.

That day, Obama's economic team had devised a possible solution. To attract enough House Democratic votes to reach a bipartisan majority of 218, they figured they needed another $400 billion of revenues—but they were willing to add extra Medicare and Medicaid savings to the package to win this additional amount. Law, Sperling, and the others were mindful not to be asking for $400 billion on its own.

Nabors, after consulting with the president, sent the proposal to Boehner's and Cantor's offices—as if he were just floating an idea of his own. The response was encouraging: Sure, let's talk about this tomorrow morning.

The next day, Nabors, Lew, Sperling, and Reed met with Barry Jackson and Steve Stombres, Cantor's chief of staff, for two hours. The atmosphere was productive. It seemed a deal was within their grasp. Each side was sensitive to what the other needed. The Republicans said they understood why the Gang of Six had made it difficult for the Obama White House. They couldn't agree to the additional $400 billion themselves—that was above their pay grade. But they did not recoil at the request. They discussed how to structure the tax reform that would generate the revenues in the deal.

Several other issues were resolved. The package would include extensions of the payroll tax cut and unemployment insurance benefits (for stimulus seekers). It would change the Social Security cost of living allowance, for about $125 billion in savings. They were still haggling over the trigger, but they were heading toward an agreement. Nabors thought they were a hair away from a deal.

Lew had to leave to attend a lunch meeting with Senate Democrats. As the meeting broke up, Jackson looked at everyone in the room and asked, "We're all trying to get to yes?"

That was the consensus.

ON CAPITOL HILL, DEMOCRATS WERE NERVOUS—AGAIN. THERE were reports that the grand bargain was alive and rumors that the latest iteration didn't contain revenues. "[When] we heard these reports of these mega-trillion-dollar cuts with no revenues, it was like Mount Vesuvius. . . . Many of us were volcanic," Senator Barbara Mikulski, a Maryland Democrat, said.

In talking with reporters, Reid issued a quasi-warning: "The president has always talked about balance . . . this can't be all cuts."

Some Democrats suspected that Daley was once more ceding territory to his pal Boehner for the sake of a deal.

Lew went into that lunch meeting and was grilled by Democratic senators anxious to know what the White House was up to. *Politico* reported that "screaming" took place. If there's a deal, why don't I know about it? Reid asked.

Why yield to the Republicans right now, the senators asked, just as Obama and the Democrats were winning over the public on the question of raising taxes to achieve shared sacrifice?

Lew assured them that revenues were not off the table. He knew that the previous night, Obama had suggested adding even more revenues, but he could not share such details with the senators. The talks were too precarious. Whatever he told the lawmakers would leak—and that could cause an uprising on Boehner's side. Lew had to absorb the venting.

Democrats were again questioning Obama's core values. Slashing the budget of federal agencies, embracing cuts to entitlements—this was too high a ransom to pay. In a meeting with policy advocates, Senator Sheldon Whitehouse, a Rhode Island Democrat, remarked, "The question the president has to ask is, is he the conciliator in chief or is he the leader of the Democratic Party?"

A variety of progressive groups, including MoveOn and the AFL-CIO, launched call-in drives targeting the offices of congressional Democrats and Obama's reelection headquarters in Chicago. "If Democrats agree to a deal that hurts working families but does nothing to make the rich and corporations pay their share, it would be a betrayal of core Democratic values and could have serious consequences for the base's involvement in next year's elections," threatened Justin Ruben, MoveOn's executive director.

Democratic strategist Donna Brazile tweeted, "What's so grand about cutting the social safety net, eliminating children's health programs, firing teachers and firefighters? That's cowardly."

A former Obama White House official expressed similar disappointment: "They keep saying, 'We're going to fight in the next round,' and they never fight."

This moment was bolstering progressive discontent. In an

interview that week, author and agitator Cornel West declared, "Poor people and working people have not been a fundamental focus of the Obama administration. That for me is not just a disappointment but a kind of betrayal." A month earlier, Al Gore had penned a piece for *Rolling Stone* expressing discouragement, observing that Obama had "failed to use the bully pulpit to make the case for bold action on climate change."

As White House aides routinely pointed out, Obama's approval ratings among self-described liberals were high; the dissatisfaction of progressive intellectuals and activists was not representative. But it was clear that if Obama conceded too much on Medicare and Social Security and didn't wring a ton of revenue out of Boehner and Cantor, he could lose Democrats on Capitol Hill and elsewhere.

Boehner also had to worry about incoming fire from his own side. He rushed to Rush Limbaugh's radio show and denied he was in league with Obama: "No deal publicly. No deal privately. There is absolutely no deal."

AFTER LEW RETURNED TO THE WHITE HOUSE, THE PRESIDENT called Boehner.

"If we don't get $400 billion more, it will be hard to get enough Democratic votes," Obama told the speaker.

The president said he understood that Boehner would have a tough time going higher on the revenues. If that wasn't possible, Obama added, they'd have to find a way to make the package more attractive (or less odious) to Democrats, perhaps by softening the entitlement provisions.

Obama was effectively saying it was Boehner's call. Revenues could go up or entitlement reductions could go down.

The president said he thought the speaker should go higher on revenues so they could reach the grand bargain.

Okay, Boehner said, I'll get back to you.

Obama and his aides now had to wait to see which course Boehner would choose.

"We all felt this was coming together," a senior administration official said.

That evening, Obama called Reid and Pelosi to the White House and told them to ready themselves and their colleagues on the Hill for a deal with assorted entitlement reductions.

WHEN SPERLING, LEW, AND NABORS ARRIVED AT THE WHITE House the next morning, they immediately asked if Boehner had called the president.

He hadn't. Obama had called Boehner but had not reached him.

When Boehner spoke to reporters that morning, he said, "Frankly, we are not close to an agreement."

At a meeting of senior staff, Obama's top aides tried to recall any previous occasion when a speaker of the House had not promptly returned a phone call from the president of the United States. No one could.

The hours went by. Nothing. Searching for an explanation, Obama's aides figured that Boehner was waiting until after the Senate vote on his Cut, Cap, and Balance bill, which had passed the House a few days earlier. But there was still no call from him, following the Senate's late-morning rejection of the measure on a 51–46 party-line vote.

"It was mind-blowing," a senior administration official recalled.

In the afternoon, Obama tried to reach Boehner, and the speaker's office said Boehner would call in about two hours. Obama asked to speak to him immediately, but Boehner would not take his call. White House aides were amazed. The speaker had disappeared for an entire day.

Sperling joked that he wouldn't want to date these guys: They just leave. They don't call to talk or say they want to see other people. They just leave.

OBABA ENTERED THE JAMMED WHITE HOUSE briefing room at 6:00 P.M. Thirty minutes ago, he said, Boehner had called him to say he was again walking out of talks for a grand bargain. This was the second time in a month that Boehner had dumped Obama.

Obama outlined what he had kicked in: a trillion dollars in discretionary spending cuts (which included reduced military expenditures) and $650 billion in entitlement cuts. And, he added, he had asked for $1.2 trillion in revenues—no hikes in tax rates, just closing loopholes, eliminating certain deductions, and tax reform.

This was, he pointed out, less revenue than the Gang of Six had proposed. It was "an extraordinarily fair deal. If it was unbalanced, it was unbalanced in the direction of not enough revenue."

"I've been left at the altar now a couple of times," Obama complained, "and I think that one of the questions that the Republican Party is going to have to ask itself is, can they say yes to anything?"

With his "balanced approach" shot, Obama was now down to his final stand: "The only bottom line that I have is that we have to extend this debt ceiling through the next election."

He told his advisers that he'd be damned if he were going through all this again anytime soon.

WITHIN AN HOUR OF OBAMA'S PRESS CONFERENCE, BOEHNER was holding his own on Capitol Hill.

"The White House moved the goal post," he claimed. He charged that "there was an agreement on some additional revenues, until yesterday when the president demanded $400 billion more, which was going to be nothing more than a tax increase on the American people." (The difference at stake—that $400 billion—represented only slightly more than 1 percent of the estimated $36 trillion to $37 trillion in government revenues for the next ten years.)

In a letter to his fellow House Republicans, Boehner declared that a deal "was never really close." He claimed that Obama had been "emphatic that taxes have to be raised" and "adamant that we cannot make fundamental changes to our entitlement programs."

None of this was true.

Obama had frightened his own party by consenting to considerable changes in Medicare and Social Security (and cuts in Medicaid), and he had accepted Boehner's proposal to produce revenues without raising tax rates.

Senior House Republican staffers, according to one, were told by GOP leadership aides that the president had kept making demands for revenues without offering specifics on entitlements. (Sperling and other White House officials would later be aghast to hear this.) Boehner and Cantor apparently had not disclosed to their own colleagues how far the negotiations had come.

WHITE HOUSE AIDES WERE LEFT SCRATCHING THEIR HEADS. They had been striving toward a deal—*after* Boehner and Cantor revived the negotiations for a grand bargain. Specifics were on the table. Paper had been going back and forth between Boehner's office and the White House. Barry Jackson had signaled he was committed to getting to yes.

Obama's aides also realized that if Boehner had returned the president's phone call and said he truly couldn't budge further on revenues, the president might well have settled for the deal as it was (without any softening of the entitlement reductions) and then done what he could to pass it in the House and Senate. White House aides wondered if they had been suckers to believe good-faith negotiations were possible with Boehner.

Later that weekend, Obama, during a phone conversation with Boehner, asked why the speaker had never called back.

"I wasn't taking anything off the table," Obama said. There was nothing he had not been willing to discuss.

Boehner offered no explanation. He was not willing to talk about it.

White House aides, after learning of the conversation, reached a simple conclusion: John Boehner couldn't deliver his own people. During that twenty-five hours of silence, he realized he couldn't sell any package of this sort to his Tea Party–dominated caucus, and there was nothing left to talk about.

HAD OBAMA INVESTED TOO MUCH IN BOEHNER AND THE ALlure of a grand bargain? "The president thought he was close to getting it done with Boehner and this animated every one of their actions," Robert Gibbs recalled.

Yet, after the fact, it was easy to argue that Boehner had never been in a position to consummate a big deal. He couldn't convince 218 of his 240 House members to pass such a package, nor could he round up enough Republican votes to give the grand bargain a fighting chance. Had he tried to bring a measure with revenues to the House floor, he might have faced an internal rebellion.

Despite Boehner's public blustering, aimed at soothing the Tea Partiers in his caucus, Obama and his aides thought the speaker had been truly interested in a historic deal and that he had not jerked them along to purposefully set up a failure, for which the president could be blamed (though that possibility did cross some minds at the White House).

Perhaps Obama and his team should have recognized sooner that Boehner could not lead his own caucus to a deal. But a crisis loomed, and they were desperate for a solution. As several White House aides later noted, it had not been their job to worry whether Boehner could corral his own members.

Plouffe and others had wanted a values debate between the president and the Republicans, but Obama had spent (wasted?) months—precious time—stuck in complicated behind-the-scenes negotiations in search of a deal that was a bridge too far.

"For months and months, we were deep in the bowels of secret negotiations," a senior White House official later lamented. "We stopped fighting them in public to get it done in private. Only it didn't work."

That was the price of Obama's effort to transcend the political and policy wars of Washington to achieve what would have been a historic compromise. An outside Obama adviser later said, "After Boehner's first walkout, it was incredible that they kept chasing that deal. They got too inside that bubble."

Obama's critics on the left—and in the middle—had frequently criticized him for lacking daring or fortitude. But it was in his nature to attempt difficult things that entail risk. Going for a big compromise had meant overcoming Republican obstinacy and potentially tearing apart his own party. It was a long shot. But the president had not been put off by the odds.

"He saw an opening and aimed for it," a White House aide observed. "Obama is self-contained and lives completely within himself. He's an introvert emotionally. But he's a risk taker. He's like an athlete who says, 'Fuck it, I'm going to shoot that half-court shot at the last minute.' He did the same on health care and bin Laden. That's how you get the big wins. He tunes out the drama. He doesn't care about, say, the relationship between Boehner and McConnell. He's able to see opportunities others don't and is willing to take these big chances. This time it didn't pan out for him."

THE NEXT WEEK WAS A BLUR OF LEGISLATIVE JOCKEYING ON CAP-itol Hill—most of it pointless. McConnell asked Obama to butt out and let the legislators fix this on their own. "Meanwhile, we were preparing for a double-dip recession," a White House official later said.

Boehner drew up a two-phase plan that called for $1.2 trillion in cuts now and $1.6 trillion the next winter, with the debt ceiling being lifted each time. Without any revenues, the second round of reductions would have to incorporate severe whacks at Medicare and Social Security. This bill was designed to appeal to Boehner's conservative wing, yet it was unclear if it was sufficiently draconian for his fellow Republicans.

On the other side of the Hill, Reid hatched his own legislation that would raise the debt limit through 2012 in return for $2.7 trillion in deficit reduction, which would include $1.2 trillion in cuts to federal agencies (including the Pentagon) and $1 trillion in savings from the end of the Iraq and Afghanistan wars. And there would be a supercommittee to work up a deficit-reduction proposal guaranteed an up-or-down vote by the end of the year.

There was not much balance or shared sacrifice in this plan. But the White House endorsed it. In a private tweet, a White House official at the time explained, "We obv want revenues, but Reid also doesn't have entitlements, so he is trying to avert a crisis. Not ideal, but better than Boehner."

EIGHT DAYS BEFORE A POSSIBLE DEFAULT, OBAMA IN A PRIME-time address from the East Room lashed out at congressional Republicans unwilling to "ask the wealthiest Americans or biggest corporations to contribute anything at all."

He blasted Boehner's bill for offering only a short-term extension, which would likely lead to a credit downgrade for the US government and trap the nation in a *Groundhog Day* situation: "The House of Representatives will once again refuse to prevent default unless the rest of us accept their cuts-only approach."

"This is no way to run the greatest country on Earth," he asserted, noting that the American people were "fed up with a town where *compromise* has become a dirty word."

Obama asked Americans to flood Congress with a simple message: "Solve this problem through compromise." He had made it clear he favored a certain policy approach. But his higher message was that he valued compromise—that he was the reasonable man in this madhouse.

Boehner responded: "The sad truth is that the president wanted a blank check six months ago, and he wants a blank check today. That is just not going to happen."

Boehner acted as if he had slept through the summer. His spin was hollow. A blank check enables *future* spending. In requesting an increase in the debt ceiling, Obama was merely guaranteeing that the US government could cover *past* spending, including the blank checks written by Republicans during the Bush years. *Politico* noted, "As if ignoring what they had negotiated—Obama agreeing to raise the eligibility age for Medicare to 67—Boehner dismissed the president as 'adamant that we cannot make fundamental changes in our entitlement programs.'"

As the speaker left the Capitol that night, a CBS News producer overheard him say, "I didn't sign up for going mano a mano with the president of the United States."

THE NEXT DAY, CAPITOL HILL PHONE LINES WERE JAMMED. A new poll showed 68 percent of Americans—including majorities across the ideological spectrum—favored compromise.

But Boehner's problems were more immediate. Numerous House Republicans were not rallying behind his debt-ceiling bill—its sharp cuts were too mild for them—and some were even scheming against it. At a closed-door meeting, Cantor told GOPers to "stop grumbling" and fall in line.

To inspire the caucus, Representative Kevin McCarthy, the Republican majority whip, showed a clip from *The Town*, a bank heist movie. In the prelude to a violent attack, a crook played by Ben Affleck says to his friend, "I need your help. I can't tell you what it is. You can never ask me about it later. And we're going to hurt some people."

His pal responds without hesitation: "Whose car are we gonna take?"

A pumped-up Representative Allen West, a full-fledged Tea Partier, exclaimed: "I'm ready to drive the car."

The Republicans were looking to smash things.

BOEHNER WAS SCRAMBLING TO PASS HIS BILL—WHICH WAS doomed to fail in the Senate. And Reid's measure would certainly be defeated in the House. But all this had to play out, almost like a predictable horror film.

At a meeting of House Republicans on July 27, Boehner had a tough message for his compatriots: "Get your ass in line."

House Democrats were calling on the president to deploy the Fourteenth Amendment—an option Jay Carney unequivocally ruled out. Washington was in chaos.

THE FOLLOWING DAY—FIVE DAYS BEFORE THE DEFAULT deadline—Obama urged congressional leaders to forget about Boehner's bill and produce a bipartisan compromise.

"We could lose our country's AAA credit rating," he remarked, "not because we didn't have the capacity to pay our bills—we do—but because we didn't have a AAA political system."

A lower credit rating, he explained, would be the functional equivalent of a tax increase; it would result in higher interest rates on mortgages, car loans, and credit cards. Stocks had been falling for six days in a row.

After much drama, Boehner brought his legislation to the House floor, and it squeaked by on a 218–210 vote. Within two hours, the Senate cast aside the bill on a 59–41 vote.

After days of frenetic legislative activity, Washington was not any closer to a resolution. Cable television networks started running countdown clocks to default.

ON SATURDAY MORNING, THE MOOD AT THE WHITE HOUSE WAS grim. Obama had tried the Biden talks. He had tried the Cabinet Room talks. He had tried direct talks with Boehner. He had tried letting Congress sort this out. But the previous night, the obvious had been demonstrated yet again: no one could get a debt-ceiling bill through both houses of Congress.

At midday, Mitch McConnell called Biden to see what they could work out. McConnell preferred negotiating with Biden over Obama. They had paved the way to the tax-cut deal of the past December. Perhaps they could find a way out of this jam.

That conversation led to a flurry of calls between the vice president and the Senate minority leader, and between their respective staffers. And there were rolling meetings at the White

House. Some were secret sessions on what the administration should do if no deal was reached by Tuesday night.

The White House and McConnell quickly agreed on the basic parameters: two sets of cuts that in total matched the hike in the debt ceiling, with up-front cuts of about $1 trillion and a supercommittee producing the second tranche of deficit reduction—or a trigger would be pulled.

Boehner was also in the mix. All through the day, he and McConnell conferred, often trudging to and from each other's Capitol Hill office, Boehner puffing away on a cigarette. The two were in their own negotiation, while McConnell haggled with the White House.

Early in these conversations, McConnell and Boehner yielded on a critical point: there would not be another debt-ceiling vote in six months. Obama had said no, and that no was sticking. The agreement would somehow extend the borrowing authority through 2012. But the talks stalled—as they had in the past— over the tricky topic of the trigger.

Then White House aides suggested sequestration— Washington-speak for automatic spending cuts. And military spending would be included. The Republicans accepted the idea. But both sides had to determine what would be subject to this budgetary doomsday device.

AS THE HOUSE PREPARED TO VOTE DOWN REID'S BILL, THE MEET-ings continued. That afternoon, Obama summoned Reid and Pelosi to the White House to tell them a deal was being hastily constructed.

Pelosi said any agreement that included a provision scaling back the cost of living adjustment for Social Security would not win Democratic votes. She also pushed for less severe cuts to Medicare providers in the trigger (from 4 percent to 2 percent).

After the meeting, a testy Reid said, "The question is, are we closer to an agreement? The answer is 'nope.'"

And Boehner was still lobbing brickbats at Obama. "It's time for the president to outline how we get out of this cul-de-sac he's driven us into," he carped during a press conference with McConnell. Boehner was sticking to the same routine of blasting Obama for supposedly doing nothing, even while he was in negotiations with the president.

COME SATURDAY NIGHT, THERE WAS NO AGREEMENT ON THE trigger. During one call late that evening between White House aides and GOP staffers, the Republicans demanded that Medicaid cuts be part of the sequestration. (If the supercommittee failed, poor people would get less health care.)

Sperling was explaining why Obama would not accept that— "there has never been a low-income program in a sequester"— when suddenly, Lew interrupted.

"No means no," he yelled. "No! No! No!"

Lew was exhausted—and also upset that the Republicans thought the White House would accept an unprecedented Medicaid sequester. In a meeting later with Obama, Sperling and Lew recounted the episode for the president.

"You did exactly the right thing," Obama said.

ON SUNDAY MORNING, THERE WAS NO IMMEDIATE WORD FROM Boehner about the deal taking shape. Dark humor set in at the White House: it's déjà vu all over again. Obama aides wondered if Boehner was trying to run out the clock as a way of forcing Obama to capitulate to an awful agreement at the last moment.

On the Sunday shows, news was spreading that a deal was near. Liberal Democratic senators, including Al Franken, Barbara

Mikulski, Carl Levin, and Tom Harkin, were pressing Reid to explain why the agreement contained no revenues. Where was the balance? they asked.

But the talks hit another snag. The deal, at this point, would establish a firewall between military and nonmilitary programs in the discretionary spending budget for two years so nonmilitary programs could not be cut deeper to spare the Pentagon. Yet House Republicans wanted fewer reductions targeting the Pentagon—and White House aides assumed this was stirring up trouble for Boehner. .

At one point, Biden was on the phone with the speaker, who was pressing the vice president to yield on this front. Biden explained that maintaining this firewall was necessary to guarantee Democratic votes. Otherwise, nonmilitary spending would be cut disproportionately to protect Pentagon expenditures.

"It's religion for Democrats," Biden explained.

But, Boehner countered, we've given you Pell grants and disaster relief funding.

"No," Biden said. "We're willing to be flexible but not broken."

"Oh, come on, Joe," Boehner said, sarcasm dripping. "The president is getting his money."

Obama and Biden refused to budge.

McConnell, too, pushed Biden on this point.

"No chance we're changing this," Biden told him.

Obama walked into Biden's office during this stretch, and the vice president told him, "I'll take this to the brink."

Don't move on the firewall, Obama said.

For a while it seemed possible that the firewall dispute could crash the whole deal. The president and his aides began pondering how to break the news to the world that once more a deal had collapsed and that default could be imminent.

Dan Pfeiffer started drafting a statement for the president to announce that no last-minute compromise could be reached.

McConnell broke the impasse by suggesting they change the placement of the firewall. Instead of separating defense and nondefense discretionary spending, it would set up a divide between "security" and "nonsecurity" spending. This would lump military operations with foreign aid, homeland security, veterans affairs, and other functions and would cushion the blow to the Pentagon.

On late Sunday afternoon, Obama convened his aides for a discussion.

"What do we think of this policywise?" he asked.

Lew quickly led a policy seminar on the possible impacts of this proposal, and Obama and his aides at the end of the conversation concluded this change would be fine.

But the Republicans wanted the security/nonsecurity distinction also applied to the sequestration in the trigger. Obama rejected that. The trigger had to be as painful as possible for the Republicans so they would be motivated to accept a second round of deficit reduction that included revenues. If the GOPers could protect the Pentagon from the full force of sequestration, they'd be less willing to sign on to revenue increases.

"In the end, the Republicans rolled over on defense," a Democratic staffer later said. "That surprised us."

Obama and Biden still had to sell the deal to the Democrats. In the late afternoon, Reid, Hoyer, Durbin, Van Hollen, and others assembled in Pelosi's office to review the possible accord. None of them was enthused about it. Biden was on the phone.

The vice president went over the technical details with Van Hollen. But his big pitch to the group was that the trigger included significant military cuts and that ought to compel Republicans to eliminate corporate loopholes to generate revenues.

Reid left to issue a statement of support. But Pelosi said she first had to discuss the agreement with her caucus. She had accomplished her mission—keeping Social Security and Medicare beneficiaries out of the range of the trigger. In next year's election, no one would be able to say (accurately) that the congressional Democrats backed Medicare benefits cuts.

AT ABOUT 8:00 P.M., OBAMA, SURROUNDED BY AIDES IN THE Oval Office, called Boehner. When Boehner got on the line, Obama started explaining the final details. But in the middle of a sentence, Boehner interrupted him.

The president's aides couldn't hear what Boehner was saying but anxiety flared.

Oh my God, Sperling thought, here we go again.

But after a few moments, Obama said, "Congratulations to you, too, John."

THE DEAL WAS DONE. IN RETURN FOR IMMEDIATE SPENDING CUTS of $917 billion spread out over ten years, Obama could raise the debt ceiling by $900 billion. (Supposedly almost half of these initial cuts would come from security spending, with the Pentagon absorbing about three-quarters of that, but the language of the actual legislation would make this a bit fuzzy.) In the second phase, a bipartisan supercommittee made up of twelve House and Senate members would have to produce by Thanksgiving a plan for $1.5 trillion in additional deficit reduction, and Congress would have to vote on the measure within a month.

If such a plan were enacted, Obama could boost the debt limit by $1.5 trillion. But if the supercommittee failed to achieve at least $1.2 trillion in deficit reduction, sequestration (divided equally between defense and nondefense programs) would kick in starting January 2013.

Medicare payments to providers could be hit by these automatic cuts, but, as Pelosi had demanded, these reductions were limited. Sequestration would not touch Medicaid, Social Security, veterans' benefits, and assorted programs servicing low-income families. And Obama could ask for up to $1.2 trillion in new borrowing authority in conjunction with the sequestration.

No matter what happened, Obama was guaranteed the debt ceiling would not again become a major legislative fracas until after the 2012 election.

The agreement had no revenues and no entitlement reductions. The supercommittee would consider both areas and squabble over all that. The deal contained no job-creation measures, but the White House inserted increased funding for Pell grants. The Republicans were guaranteed votes in the House and the Senate on a balanced budget amendment.

AT 8:40 P.M., JULY 31, 2011, OBAMA ENTERED THE WHITE HOUSE briefing room to announce the deal. The result, he noted, would be a historically low level of spending that still allowed "job-creating investments in things like education and research." He had made the same argument at the end of the budget fight.

"Is this the deal I would have preferred?" he asked. "No." But the agreement "will allow us to avoid default and end the crisis that Washington imposed on the rest of America."

Obama, once again, was blaming Washington more than Republican extremism. Even at this late stage, Obama did not feel free to lash out. The debt-ceiling deal still had to move through the House, and if Obama were to lambaste the GOP, he could diminish the prospects for passage.

This agreement, Obama remarked, "will allow us to turn to the very important business of doing everything we can to create jobs, boost wages, and grow this economy . . . and that's what we should be devoting all of our time to accomplishing in the months ahead."

Biden tweeted, "Compromise makes a comeback."

In a conference call late on Sunday, Boehner told House Republicans, "I'm not celebrating."

But as of this moment, the plan was all cuts—no revenues, no shared sacrifice, no balance. He told his fellow Republicans, "This isn't the greatest deal in the world, [but] it shows how much we've changed the terms of the debate in this town."

That much was true.

Obama had ended up with a small cross. He had averted a default that never should have been possible. But he had paid Boehner ransom—though some of that payment was coming out of the Pentagon's hide.

The president had beat back the GOP attempt to pin him down with repeats of the debt-ceiling debacle prior to the 2012 elections, but he had not been able to prevent the establishment of a troubling precedent: the debt ceiling could be a bargaining chip. He had protected certain funding priorities, but he had not won a package with job-creating measures. He had persuaded the public that the Republicans' cuts-only stance was wrong, but he had not found a way to translate that into political leverage he could use to influence the outcome. He had looked reasonable, but he had ended up the lead player in an unpopular Washington fiasco.

"Not getting a second debt-limit vote was what we cared about," a senior administration official said. "They were getting $1 trillion in discretionary spending cuts this first time. The next time, they'd come back and demand $500 billion out of Medicare and Social Security. Imagine if Pelosi had said to President

Bush, when she got the speaker's gavel, 'You pull out of Iraq or we blow up the global financial system.' If we gave them a second vote, who knows where this would go?"

Liberal Democrats were appalled. Obama may have blocked irresponsible Republican extremists from causing a default, but for them he had bartered away too much.

Representative Raúl Grijalva, the head of the Congressional Progressive Caucus, proclaimed, "This deal trades people's livelihoods for the votes of a few unappeasable right-wing radicals, and I will not support it."

Representative Emanuel Cleaver, who chaired the Congressional Black Caucus, dubbed the accord "a sugar-coated Satan sandwich." On MSNBC, he complained, "I came back to Washington at the beginning of the year thinking we were going to create jobs, and we allowed the national discourse to change from jobs to the debt. . . . If I were a Republican, I would be dancing in the streets."

Under the headline "The President Surrenders," Paul Krugman, in his *New York Times* column, contended the deal was "a disaster, and not just for President Obama and his party. It will damage an already depressed economy; it will probably make America's long-run deficit problem worse, not better; and most important, by demonstrating that raw extortion works and carries no political cost, it will take America a long way down the road to banana-republic status."

THE NEXT NIGHT, THE QUICKLY DRAFTED BILL WAS BROUGHT TO the House floor. Pelosi intended to vote yes—to avoid a possible financial crisis—but she did not press her members to follow her lead. This was a bitter moment for her. The Tea Partiers, she sadly realized, had succeeded in shifting the Washington conversation toward an obsession with deficits. "They did win that," she said later. "They changed the arena." And the debt-ceiling crisis, as she saw it, was part of their blueprint to demolish government with spending cuts targeting crucial programs: "Clean air, clean water, education—they will destroy it."

When the voting started, the Democrats held back as the Republicans cast votes. "Let them put 218 on the board," Pelosi told her party colleagues, "and then we'll see."

As the minutes ticked by, it became evident that the Republicans couldn't reach a majority. About a quarter of Boehner's caucus was voting against him. Once again, the speaker couldn't deliver.

With a few minutes remaining, the Democrats, having made their point, started voting. The final vote count was 269–161. Half of the Democrats voted for the Satan sandwich. Boehner lost 66 Republican votes. The ideologues of his party, who had prevented him from achieving the grand bargain and who had pulled the debt-ceiling deliberations far to the right, were still off the reservation.

There was little celebration in the chamber. Yet one bright moment occurred when Representative Gabrielle Giffords, on her first visit to the House since surviving that assassination attempt seven months earlier, received a standing ovation as she cast a vote for the bill.

The next day, the Senate passed the compromise with little drama—hours before the midnight deadline.

THE PREDICTABLE DEBATE KICKED OFF: COULD OBAMA HAVE handled this better? Liberal commentators asked why Obama hadn't slammed the Republicans from the start, issuing clear demands—no chaining of the debt ceiling to deficit reduction, no deficit reduction without higher taxes on the wealthy—and sticking with them.

One outside evaluation was particularly stinging. Jared Bernstein, who had recently been Biden's chief economic aide, noted that because the president did "not have a strategy to counteract [Republican] extremism," he had been forced into the position of accepting "a plan *far less balanced* than" he would have liked: "This was an ugly debate where reckless ideologues got the better of the grown-ups in the room who were not willing to risk the economy to protect the government."

The bad review from commentators, naturally, irritated White House aides, who believed the president had demonstrated courage (by striving for the grand bargain) and responsibility (by eschewing the routine brawling of politics to avert a catastrophe).

"Our own pundits were decrying us for dealing with hostage takers, as if there was another option," a senior administration official griped.

White House aides even felt a bit heroic. They had saved the country. Had the United States stopped paying its bills, the whole global financial house of cards—see Greece—could tumble.

As the debris was settling, Axelrod maintained that the deal had "removed an obstacle" for Obama. Now the president could finally return to his growth agenda and the push for near-term job creation.

"We've lifted a brick off that," Axelrod said.

What about the narrative that Obama was forced to cave in the face of Republican intransigence?

"I don't think the American people give a damn about that," Axelrod contended. "The impression people are getting is that the Republicans are willing to tank the economy to get their ideological prerogatives and to beat the president. This is not a situation where we got 'em just where we want 'em. But who put themselves in a better position to win later? We did better than they did."

A senior administration official echoed this point: "We

definitely didn't win this. We took a hit and our share of the blame for the dysfunction. But the Republicans seriously soiled themselves."

To answer its critics, a defensive White House released a long fact sheet on "myths and facts" about the agreement that insisted the deal was a "a win for *all* Americans." The first myth it cited was "President Obama caved." No, he had avoided default, protected entitlements, and set up a path that would pressure Congress to adopt a balanced approach.

The White House was again selling an unpopular compromise with benefits that were largely based on preventing negatives from happening: we stopped the crazies from ruining the economy. The December tax-cut compromise had included provisions to assist Americans confronting economic troubles; it had the potential to boost the economy. Democratic and independent voters, according to polls, had reacted favorably to that compromise. The debt-ceiling deal had few, if any, explicit policy advances that could juice the economy in the short run.

The immediate polls found that a majority didn't like the deal. And a CNN/Opinion Research Corporation survey noted that only 46 percent approved of Obama's handling of the debt negotiations. But it was worse for the Republicans: 14 percent responded positively when asked how Congress had been doing its job. In a *USA Today*/Gallup survey, half of independents—the target group for Obama's be-the-adult-and-compromise strategy—were unhappy with the compromise; only 33 percent approved.

But there was a sliver of good news for the White House. By the end of the messy episode, a significant majority of Americans, according to numerous polls, backed Obama's overall approach to deficit reduction over the GOP's slash-slash-slash strategy.

This suggested that when Obama presented a clear vision of his values and priorities—and contrasted them with those of the Tea Party–fueled Republicans—he could gain ground. This was consolation for Obama and his team as they looked ahead to future confrontations and, of course, the reelection campaign.

The Republicans absorbed a different lesson. McConnell boasted that this "creates an entirely new template for raising the national debt limit." And he called the debt ceiling "a hostage worth ransoming"—as if he was enthusiastic about a rerun down the road.

Representative Kevin McCarthy expressed delight that the rigidity of the House GOP freshmen had bolstered his side, saying, "You had a fear of how far they would go. I'm sure the president looks back, too, and was fearful. But in negotiations, isn't that the best thing?"

After the crisis was over, Vice President Biden shared his own theory with his aides. He wished they had let the Republicans shut down the government in April. The Tea Party was obviously looking for a dramatic moment. They didn't get one with the 2011 budget deal—and they ended up feeling cheated on the numbers, due to Lew's brilliant handiwork. But after that, the House Republicans were brimming with suspicion and loaded for bear.

That meant the Obama White House had been negotiating with a party dominated by a group looking to act out. A government shutdown would probably have been less damaging than a default. Perhaps it would have been best, Biden pondered, to have the big battle earlier—and on a smaller cross.

THE STOCK MARKET SLUMPED IN THE DAYS FOLLOWING THE deal. The episode had not bolstered confidence. Standard &

Poor's for the first time downgraded the nation's credit rating, explaining that the "political brinkmanship" was a sign that "America's governance and policymaking" were "becoming less stable, less effective, and less predictable."

Soon after the deal, economic activity, as measured in the Business Outlook Survey conducted by the Federal Reserve Bank of Philadelphia, hit its lowest point since March 2009. With the nation having reached the precipice of default, it seemed the whole economy was standing still—and Washington was broken.

THE COLLAPSE OF THE GRAND BARGAIN WAS A CRITICAL MOMENT in the Obama presidency. A former senior Obama administration official noted that at the White House—and in the Oval Office—the realization sunk in that even if the president wanted to take risks to compromise with the Republicans, he couldn't succeed: "There was no pot of gold at the end of the adult-in-the-room rainbow, and there were only so many times Lucy could pull away the football."

Obama had been struck by the Republicans' willingness to damage the economy and to tilt the playing field even further toward the wealthy. He had assumed—perhaps naively—that during an economic crisis, the Republican leadership would do something to help. And he and his aides had felt obliged to find that common patch of ground, though it might have been small and distant.

"This is the fundamental insight," a senior administration official said. "What made this year so difficult was not just divided government or how obstructionist the Republicans have been. What was so challenging was that we had to face this degree of division at a time when our economy demanded we take responsible action to prevent a downturn or a sustained period of very high unemployment. That is a major constraint. It was not that everybody in the White House could not figure out how to have good political positions. We can.

"Our obstacle to hitting the most ideal message was simply that our paramount responsibility is to do the right thing for the US economy."

The debt-ceiling fiasco had, at least, been instructive, Axelrod believed: "People are not rooting for gridlock and partisan strife—at least the vast majority are not. But the folks on the other side fundamentally believe cooperation is akin to treason. In that set of circumstances, you have to push back harder. The public recognition of that is there now."

He added that the path forward was clear: "We're in a new period. We need to confront them."

If compromise was not possible, Obama would have no alternative but to engage in open combat with the Republicans—and, with the GOPers no longer holding the economy hostage, he would have the freedom to do so.

As a senior administration official noted, "We don't have a gun to our head anymore."

THE DAY OBAMA SIGNED THE DEBT-CEILING BILL INTO LAW, he held a previously scheduled meeting in the East Room with the AFL-CIO's executive council. He was eager to explain himself.

He told the unionists he had had no other choice but to cut this deal to remove the debt-ceiling "millstone around our necks." He had shielded entitlement and low-income programs from the GOP hatchet. And now he was in a better position to advocate for job-creating programs. To have credibility for that crusade, he argued, progressives had to show they can manage the economy responsibly, and, like it or not, the American people were genuinely concerned about government debt. That could not be ignored.

He counseled the unionists not to be discouraged by the dispiriting debt-ceiling fight.

"I won the argument," he said. "I wasn't willing to burn down the house or split the baby."

And he maintained that the administration had been confronted with impossible circumstances: "It's not like if I had been a better negotiator, we'd all be better off."

"I'm attacked every day for my socialist tendencies," Obama said, with a touch of sarcasm. But he vowed, "I'm not through fighting."

He reiterated the message he had conveyed to the union presidents the previous December: This is a long fight; the House Republicans' agenda started with Barry Goldwater in 1964. "They got whipped but stayed with it. . . . Now everyone is mopey and mad at me. Don't doubt my commitment.

"Don't doubt who I am fighting for," he added. "The other side knows and that's why they are trying to beat me."

He promised the labor officials that "the only thing we will talk about for the next nine to ten months will be how to grow the economy and put people back to work."

Toward the meeting's end, he remarked, "Given the economy, it's remarkable that we are still in the game."

Chapter Thirteen

OBAMA UNLEASHED

On the last day of August 2011, President Obama stood at a podium in the Rose Garden in the midmorning sunshine.

Standing beside him were David Chavern, executive vice president of the Chamber of Commerce, and Richard Trumka, president of the AFL-CIO. The two men didn't say anything; they were props, there to help Obama push a bill stalled in Congress that would provide funds for highway construction, bridge repairs, mass transit, and other projects. This was part of the president's newly revitalized—post-debt-ceiling—focus on jobs and more jobs. In a week's time, Obama would deliver a speech to Congress outlining a jobs package.

After finishing his remarks, Obama grabbed Trumka and pulled him into the Oval Office for an unscheduled chat. No staff, Obama said to the labor leader.

Days earlier, Trumka had been the guest at an on-the-record breakfast with reporters. Referring to a meeting he'd recently had at the White House, Trumka told the journalists he had fervently urged Obama to draft a comprehensive jobs blueprint, not a collection of watered-down half measures designed to pass in the Republican-dominated Congress. If Obama in his much-anticipated jobs speech did not "propose bold solutions on the jobs crisis," and was only "nibbling around the edge," Trumka huffed, "history will judge him and I think working people will judge him."

Accusing Obama of limiting his proposals to "those little things that he thinks others will immediately accept," Trumka was essentially calling the president a small and spineless leader. And he tossed out a threat: if the president came up short, the AFL-CIO might be better off spending its money on projects other than Obama's reelection in 2012.

The president didn't appreciate preemptive potshots or speculative threats from his allies. He didn't like being doubted—which he had been for months during the debt-ceiling debate and the near government shutdown.

In the Oval Office, he chewed out Trumka.

You're losing our guys, Trumka countered.

Don't blast me in public, the president said. Obama had told Trumka and his union buddies to have faith in him—and he had meant that.

After Trumka left the Oval Office, he didn't say much about the meeting. But the next week—three days before Obama was to unveil his jobs initiative—Trumka was with the president in Detroit for a Labor Day rally. Speaking to the upbeat crowd of thousands before Obama appeared, Trumka hailed the president as "the man who worked with auto workers to save America's auto industry. That's the kind of bold, courageous action that we need right now."

And Obama was planning to be bold again that week.

A MONTH EARLIER, AS SOON AS OBAMA WAS CLEAR OF THE DEBT-ceiling tar pit, he had ordered Gene Sperling to bring him a major jobs plan by the next week. Be bold, the president said. And be credible.

Sperling had already been preparing for this. Through the spring and summer, as the president and his aides watched the job-growth numbers plummet, they had realized that as soon as they could extricate themselves from prevent-catastrophe negotiations they would have to do something major about jobs.

In mid-June, Bill Daley had convened a senior White House staff retreat at nearby Fort McNair, and he reserved a stretch of time for Sperling to explain possible jobs proposals the president could soon unveil. At the top of his list was extending the payroll tax cut and unemployment insurance benefits, which were each set to expire at the end of the year. At the retreat and in other meetings, White House aides had reminded Sperling that he had claimed that the payroll tax cut, the big-ticket item in December's tax-cut compromise, would provide a big boost. So why was the economy still sputtering?

Sperling answered that the payroll tax cut had cushioned the blow caused by the steep rise in gas prices. He was again stuck with the economic team's mantra: "It would have been worse. . . ."

In August, Sperling and his economic squad crunched numbers and evaluated options. With the economy still in a slump and perhaps heading toward a turn for the worse, Sperling estimated that the jobs package had to be at least $350 billion to have a shot at lowering unemployment significantly and pushing growth higher than 3 percent.

David Plouffe and Dan Pfeiffer were not happy about this large price tag. Yet they accepted the economic reality at hand. You couldn't move the sluggish economy without a sizable kick. Obama, too, saw that.

White House staffers held numerous meetings—with and without Obama—as they pondered assorted provisions Sperling had lined up: continuing the payroll tax cut and expanding it to cover the amount paid by employers; funding an infrastructure bank; developing a school modernization program. Everyone agreed that the individual parts of the plan should have attracted bipartisan support in the past and should be able to stand on their own, if necessary.

In one of the meetings, Obama asked Sperling, "Put Congress and politics aside. If you were trying to fix the economy what would you do?"

Spend money to help states hire and retain teachers, Sperling said. This would help preserve consumer demand in the short term and prevent the education system from deteriorating and becoming a drag on the nation's future prospects.

Why isn't that in the plan? Obama asked.

The Republicans won't pass it, Sperling said.

Let's decide what we believe is best, Obama said. He soon after agreed to support a major teachers provision—and Biden suggested adding police officers, firefighters, and first responders to this part of the under-construction package.

In other meetings—and in e-mails and phone calls—Obama requested that Sperling and the economic team increase the amount of money for infrastructure projects and incorporate a summer jobs program. He thought his policy advisers were being hemmed in by the politics of the possible and pushed them toward a bigger plan.

At the same time, Sperling and the economic team were concocting a specific and comprehensive deficit-reduction blueprint that Obama would release after unveiling the jobs bill. It would essentially be the president's submission to the supercommittee that would be struggling to cook up Part II of the debt-ceiling settlement.

The main issue for White House aides was whether the president's deficit plan should include the concessions Obama had offered during his doomed attempt at a grand bargain.

Some argued the president was already half pregnant when it came to raising the Medicare eligibility age and altering the Social Security COLA. If he included these policy changes in the plan, he would demonstrate (once again) seriousness to the deficit hawks of Washington *and* please the ever-important markets. Others maintained that embracing these measures without extracting anything from the other side was a poor negotiating strategy. And this was no longer a negotiation.

WITH THE JOBS PLAN COMING TOGETHER, OBAMA AND HIS AIDES considered the timing of its release and whether or not to rush it out. The economy seemed to be tanking. Some economists were telling the White House that there was a fifty-fifty chance of another recession.

It was mid-August, and later in the month the president would be heading to Martha's Vineyard for a family vacation. Geithner and others asked how Obama could go a few weeks without taking decisive action. This was a bad photo-op waiting to happen.

But Plouffe was determined that Obama's job-creating initiative penetrate the public. To achieve that, the White House needed to mount a sustained effort over time, with various events and appearances.

There was simply not enough runway in August to launch what was now the most important project of the Obama presidency. If Obama couldn't convince the public he was sweating for jobs, jobs, jobs, not much else would matter. And who in Washington or elsewhere pays attention to anything in the middle of the summer? It would be an act of reckless negligence to kick off an economy-saving (and presidency-saving) endeavor in the dog days.

"The world never rewards political incompetence," Plouffe argued.

Pfeiffer suggested they announce that after Labor Day the president would deliver a major economic speech unveiling a jobs plan. There would in the meantime be leaks about the initiative and the internal discussions. But that was not too high a price to pay for conveying the message that the president was back on the jobs beat.

On August 13, the *New York Times* published an attention-grabbing front-page story claiming that the White House was engulfed by a civil war, divided between aides who favored modest and low-impact economic ideas that could pass Congress (Plouffe and Daley) and those aides (Sperling and his crew) who advocated bigger, pie-in-the-sky ideas.

The article was wrong—there had not been major disagreement over the package—but it set off tremors, especially among Democrats and progressives. Here we go again, they thought: Obama was going to punk out.

The night the *Times* story hit, Pfeiffer called one of the two journalists who wrote the story. "You're going to be embarrassed," he said.

Days later, as Obama was on a campaign-style bus tour in the Midwest—blasting Congress for doing nothing to create jobs—the president disclosed that he would announce in September a "very specific plan" to create jobs. And if lawmakers wouldn't act, he added, "we'll be running against a Congress that's not doing anything for the American people."

That week, his Gallup approval ratings sank below 40 percent for the first time—and only 26 percent had faith in his handling of the economy. Congress's approval rating, however, fell to 13 percent.

AFTER OBAMA AND HIS FAMILY RETURNED TO WASHINGTON from their vacation, the jobs plan had to be finalized. The tab for the package had crept up to $375 billion, and Sperling and others were considering what provisions to shave. Then on September 2, the new monthly jobs report showed no jobs growth at all the preceding month. Zero. Upon hearing the news, Sperling felt as if he had been punched in the stomach.

Obama's advisers feared a deep economic crunch was at hand. Sperling wondered if $350 billion had been the right amount before, how could an amount near that be the appropriate level now?

He decided they should go with everything under consideration—no shaving—which would hike the cost to $450 billion. In White House meetings, he said such an amount was necessary to hit the 1.5 million jobs target. If we're serious, he asserted, the package must grow.

He knew this would cause unease among White House aides. At this size, the jobs plan would be more than half the cost of the much-maligned stimulus. That had no policy significance, but it could be symbolic ammo for the Republicans. And with Obama absolutely committed to paying for the package, the bigger it grew, the more difficulty the White House would face in tapping revenues or savings to cover the tab.

At an Oval Office meeting days before the jobs speech, Geithner, who was usually Mr. Fiscal Discipline, backed up Sperling. Plouffe said the larger price tag would make the plan harder to sell, but noted that anyone inclined to blast it would do so whatever the size. Plouffe didn't try to knock down Sperling's case.

The final number was Obama's call. "If you go to Obama and say this is the right thing to do economically, he doesn't like to say no," a senior administration official remarked. "Maybe that's his downfall."

The president agreed with Sperling. His bold package would become $75 billion bolder.

As Obama was poised to buoy his Democratic and progressive base—and demonstrate to independent voters he was fixated on jobs—he enraged environmentalists and liberals by killing a proposed EPA standard for ground-level ozone, pollution caused by power plants, industrial facilities, and vehicles, commonly known as smog.

This rule would have forced states and cities to reduce local air pollution or be hit with federal penalties. (Smog poses a serious health risk to children, seniors, and people with chronic lung diseases.) House Republicans, led by Eric Cantor, and industry groups had been on the warpath about this pending regulation for months.

In a statement, Obama noted that his EPA had issued other standards that had removed billions of tons of pollution from the nation's air, but he said this particular rule should be yanked because it would be up for reconsideration in two years. The president maintained that it would be unreasonable to ask state and local governments to implement a new standard that could soon change. This was part of his ongoing effort, Obama said, to reduce "regulatory burdens and regulatory uncertainty."

Republicans had been claiming that the administration's excessive regulation was a primary cause of the nation's economic woes—though the Obama administration had issued fewer regulations than the Bush administration. The GOP had no data to back up the charge that these regulations—as opposed to a severe lack in demand—were impeding businesses. Still, as Congress returned to town at the end of the summer, Cantor sent out a memo outlining the House Republicans' two-part jobs agenda. The first priority was repealing "jobs-destroying regulations"; the second was cutting taxes.

Earlier in the year, Obama had signed an executive order

calling for a government-wide assessment of rules in search of outdated regulations that stifle businesses. As this regulatory review proceeded, a former senior Obama White House aide noted, "If Obama wants anyone in the business community to believe he's serious about regulatory reform, he's going to have to make an example of a regulation people on his side like."

Whether or not that was Obama's intention, the smog case fit the bill. His decision to dump the new smog standard came as a shock to environmentalists. This was a complete win for industry. White House aides claimed that the smog decision did not signal a wider surrender on environmental regulation—and assured environmentalists that Obama still stood behind administration efforts to bolster standards governing air toxins. (In December, Obama's administration would finalize tough new curbs on mercury and other poisons emitted by coal-fired utilities.)

For some liberals, the death of the smog safeguard was another indication that Obama was too willing to sell out. "I have no idea what Barack Obama—and by extension the party he leads—believes on virtually any issue," Drew Westen, a psychology professor and messaging consultant to Democrats and progressive groups, had written in a *New York Times* op-ed slamming Obama as a weak and temporizing leader days after the debt-ceiling battle. Yet the smog decision had come just as Obama was preparing to take one of the more progressive strides of his presidency.

THE FIRST APPLAUSE LINE IN OBAMA'S JOBS SPEECH, DELIVERED to a joint session of Congress on the evening of September 8, 2011, referred to the debt-ceiling madness of recent weeks: "The question is whether, in the face of an ongoing national crisis, we can stop the political circus and actually do something to help the economy."

Senators and House members clapped; some cheered. Obama's intent was clear: it was time to move on.

"We felt liberated," a senior administration official said. "Through the summer, we were dealing with House Republicans who were both irresponsible and incompetent. It was like you handed a nuclear bomb to inept and useless people. They couldn't count votes. They didn't know what would pass. And anything we said could blow everything up."

Obama went on to describe the American Jobs Act that Sperling had assembled for him. To help small businesses hire and grow, Obama proposed cutting the employer portion of the payroll tax (and eliminating it for new hires) and extending tax breaks for businesses that build new plants and purchase equipment. The payroll tax holiday for workers enacted the previous year would be extended and expanded, which would mean an extra $1,500 for a family earning $50,000. Unemployment benefits would be extended, and tax credits offered to employers who hire out-of-work veterans. States would receive $35 billion to prevent layoffs of hundreds of thousands of teachers, firefighters, and police officers.

The president proposed investing $30 billion in school infrastructure to modernize at least 35,000 public schools, with some of this money underwriting science labs and Internet-ready classrooms. There would be $50 billion to fund highway, transit, rail, and aviation projects; $25 billion for rehabilitating and refurbishing vacant and foreclosed homes and businesses; and $10 billion to capitalize an infrastructure bank. And there was more: tax credits for hiring long-term unemployed workers, a youth summer jobs program, and expanding access to high-speed wireless services in remote rural communities.

The plan offered tax cuts, which conservatives could like, and noncontroversial stimulus projects—though Obama didn't use that word. In a few days, Obama would call for financing the plan by limiting itemized deductions for individuals earning more than $200,000, ending subsidies for oil and gas companies, and changing the depreciation rules for corporate airplanes. (Senate Democrats—some weren't fans of killing oil firm subsidies or capping charitable deductions—would eventually replace these "pay-fors" with a 5.6 percent surtax on millionaires, sure to irritate Republicans.) The total cost: $447 billion.

In his speech, Obama noted that most of these provisions had previously been supported by Republicans. Over a dozen times, he urged Congress—meaning the Republicans—to pass the bill. Referring to the pressing need to rebuild America, he cited a "bridge that needs repair between Ohio and Kentucky"— that is, a project important to Boehner's congressional district and McConnell's home state.

He tried to box in the Republicans on the payroll tax cut: "I know that some of you have sworn an oath to never raise any taxes on anyone for as long as you live. Now is not the time to carve out an exception and raise middle-class taxes."

Perhaps the most powerful moment came when the president voiced a populist-tinged sentiment: "Should we keep tax loopholes for oil companies? Or should we use that money to give small business owners a tax credit when they hire new workers? Because we can't afford to do both. Should we keep tax breaks for millionaires and billionaires? Or should we put teachers back to work so our kids can graduate ready for college and good jobs? This isn't political grandstanding. This isn't class warfare. This is simple math."

And this was not an angry populism of class resentment. It was, as could be expected with Obama, a calm populism: there's a rational choice to be made, there's only so much money—do we protect the pocketbooks of the well-to-do or invest more in education?

The president was delivering a passionate defense of government—a perspective that had been overwhelmed throughout the year by the Tea Party invasion of Washington and the subsequent negotiations, shouting, and headlines concerning spending cuts, deficit reduction, and the national debt.

At a time of economic challenge, Obama said, the "notion that the only thing we can do to restore prosperity is just dismantle government, refund everybody's money, and let everyone write their own rules, and tell everyone they're on their own—that's not who we are. That's not the story of America."

True, Americans are rugged individualists and self-reliant, but "there's always been another thread running throughout our history—a belief that we're all connected, and that there are some things we can only do together, as a nation."

This was not a speech to persuade the audience sitting before him. The Tea Partiers would not be moved. But as he had done with the anti-Ryan speech in April, Obama was setting up the counternarrative to the Republican claim that government was the problem and had to be gutted for the economy to rebound.

His jobs plan was part of this overarching story—which he would be telling for the next fourteen months, until he faced the voters once again. And he was picking a fight: "This plan is the right thing to do right now. You should pass it. And I intend to take that message to every corner of this country."

Here was an unfettered Obama who had moved from the negotiating table to the boxing ring. His aides insisted this was not a politically calculated pivot; it was driven by circumstances. He could no longer work the inside game with Republican leaders who were held hostage by the extreme ranks of their own party; he had to go to the outside and apply pressure. He had not needed to be talked into this strategic shift, his aides recalled.

"You will see more energy and more passion," a top Obama strategist remarked. "He feels it."

WITH POLLS SHOWING THE PUBLIC DISGUSTED WITH CONGRESS, John Boehner responded to Obama's speech cautiously and in a noncombative manner, noting, "The proposals the president outlined tonight merit consideration."

Eric Cantor was also respectful: "It's time for Washington to come together and produce results." But he added that Obama's intent to pressure Congress to pass the package in its entirety was not "the right approach." Reince Priebus, chair of the Republican National Committee, was not shy about telegraphing the inevitable GOP attack: "Despite one failed stimulus, the president wants even more deficit spending."

Democrats on Capitol Hill tended to cheer now that Obama was on the offensive. "A light has gone on at the White House," a senior House Democratic aide said. "Any expectation that you can establish bona fides by governing with these Republicans is gone."

There was, however, the to-be-expected moaning. "Senate Democrats were ticked off at the White House for not consulting with them sufficiently on the composition of the bill, and running against a do-nothing Congress, rather than running against do-nothing and crazy Republicans, thus lumping Senate Dems in with the Republicans," a senior Democratic Senate aide remarked.

In pundit land, the hard-to-please Paul Krugman wrote, "It's much bolder and better than I expected. President Obama's hair may not be on fire, but it's definitely smoking; clearly and gratifyingly, he does grasp how desperate the jobs situation is. . . . Mr. Obama may finally have set the stage for a political debate about job creation."

The plan also passed establishment muster. Moody's Analytics predicted Obama's proposal would add 2 percent to GDP growth, create almost two million jobs, and lower the unemployment rate by 1 percent.

Obama hit the road to promote the jobs bill. After the midterm elections, he and his aides had concluded that in the first two years the president had not escaped the Beltway confines often enough. They had resolved for him to get out more. But the Arab Spring, budget negotiations, and the debt-ceiling quarrel had forced him to stay in Washington. At last, he was free to roam.

The first stop was the University of Richmond, which happened to be in the swing state of Virginia and also in Cantor's congressional district. (Just a coincidence, White House aides insisted.) At a rambunctious rally, Obama proclaimed, "There are millions of unemployed construction workers across America ready to put on their tool belt and get dirty. I don't know about you—I don't want the newest airports, the fastest railroads, to be built in China. . . . I want them to be built right here in the United States of America."

The crowd chanted, "USA, USA, USA." A speech focused on deficit reduction would not have elicited such an outburst of patriotism.

Four days later, the president was at Fort Hayes High School in Columbus, Ohio—not too far from Boehner's district (again, coincidentally)—for another rally. "Budget cuts are forcing superintendents . . . all over the state to make [teacher] layoffs they don't want to make," Obama declared, adding, "Tell Congress to pass the American Jobs Act so we can put our teachers back in the classroom."

The next day, in Raleigh, North Carolina, Obama said there

were 153 structurally deficient bridges in the state. If Congress passed his jobs plan, money would be available to repair many of them, and the typical working family in North Carolina would pocket a $1,300 tax cut.

Obama was finally doing what some of his Democratic and progressive allies had long urged: taking his case to the American public in a sustained fashion, applying pressure on obstructionist Republicans, and championing policies that addressed immediate economic concerns.

In the coming weeks, he would take his "pass this bill" tour to Denver, Colorado; Mesquite, Texas; Pittsburgh, Pennsylvania; Asheville, North Carolina; Hampton, Virginia; Manchester, New Hampshire; Scranton, Pennsylvania; and elsewhere—with practically each stop in a battleground state that would be crucial to Obama's electoral fortunes the following year.

Obama's conservative critics were right—he was beginning the campaign of 2012.

OBAMA HAD SHIFTED TO A JOBS-FIRST MESSAGE. BUT WITH THE supercommittee set up by the debt-ceiling deal kicking into gear, he still had to tend to deficit reduction. The recent fight had resolved little, and the policy and political reasons that had driven Obama to accept deficit reduction as a priority remained. So once again he had to put forward a plan.

Almost two weeks after the jobs speech, Obama entered the Rose Garden to announce a detailed road map for smiting deficits.

The blueprint was basic: $3.2 trillion in deficit reduction over ten years, with $1.5 trillion to come from heftier taxes on the wealthy. Obama would pull the plug on the top-bracket Bush tax cuts *and* limit tax breaks for those making more than $250,000. He would close loopholes for special interests, such as oil companies, and his plan would adhere to the "Buffett Rule"—named after billionaire investor Warren Buffett—ensuring that people making more than $1 million a year would not pay a smaller share of their income in taxes than middle-class families.

On the cutting side, Obama would slice $250 billion out of excessive Medicare payments (not benefits) and the same amount from assorted non-health-care mandatory programs (such as farm subsidies). His plan pocketed the $1 trillion savings associated with the wind-down of the Iraq and Afghanistan wars.

Obama had tossed aside the concessions he had offered Boehner weeks earlier. And he was giving up on his relationship with Boehner—which had not paid off. The president slammed the speaker for having recently proclaimed that the supercommittee could only contemplate spending cuts, not revenues: "The speaker says we can't have it 'my way or the highway,' and then basically says, 'my way—or the highway.' "

Worse, Obama noted, Boehner was putting the nation at a disadvantage: "That means slashing education, surrendering the research necessary to keep America's technological edge in the 21st century, and allowing our critical public assets like highways and bridges and airports to get worse. It would cripple our competitiveness and our ability to win the jobs of the future."

This set up the predictable showdown with Republicans. That day, Boehner told Fox Business Network, "I don't think I would describe class warfare as leadership."

Republicans started whining that Obama's plan was the "largest tax hike in US history."

It did seem at last that Obama was in a fair fight, no longer hamstrung by the burdensome responsibility of preventing Tea Party Republicans from shutting down the government or causing a financial crisis. *New Yorker* writer Ryan Lizza zapped out a half-snarky tweet: "This may be the first time in a major

negotiation with Repubs where Obama didn't give away everything in his opening bid." The *New York Times* reported that
Obama was in a "new, more combative phase of his presidency."

THIS TIME OBAMA DID NOT THROTTLE BACK, AS HE HAD AFTER
the anti-Ryan speech in April. At a fund-raiser in Manhattan, he
dismissed the "moans and groans from the other side about how
we are engaging in class warfare and we're being too populist."

In Cincinnati, Ohio, Obama stood before the Brent Spence
Bridge—the one he mentioned in his jobs speech—and noted
that this public work, which had been declared functionally ob-
solete, "just so happens to connect the state that's home to the
speaker of the House with the state that's home to the minority
leader of the Senate. Sheer coincidence, of course."

At the annual convention of the Congressional Black Caucus,
Obama wagged his finger at allies who might have become discouraged with him: "I don't have time to complain. I am going
to press on. . . . Take off your bedroom slippers, put on your
marching shoes. Shake it off. Stop complaining, stop grumbling,
stop crying. We're going to press on. We've got work to do."

Obama was not put off by the continuing accusations of class
warfare. David Brooks, the White House's favorite conservative
columnist, had claimed that Obama's "populist cries" will "fire
up liberals but are designed to enrage moderates." (This was
an absurd charge—that Obama and Plouffe were purposefully
trying to anger independents.)

Mark Penn, who had been Hillary Clinton's strategist and
pollster in 2008, warned that Obama, with his "class warfare,"
was "abandoning" the center. Yet at a Democratic Party fund-
raiser in Seattle, Obama proclaimed, "If asking a billionaire to
pay the same tax rate as a plumber makes me a warrior for the
middle class, I'll wear that as a badge of honor."

Obama was no longer seeking to be the compromiser in chief and the reasonable adult in Washington's unruly sandbox. That might have once been an effective strategy for appealing to independents. But it made no sense if pursuing bipartisan collaboration resulted in Obama looking ineffectual. "The only option was to take our argument to the country," a senior administration official said.

But Obama and his aides were not giving up on their effort to win back independents. They believed the president had attempted mightily to govern in a bipartisan manner and had been rebuffed—and that many independents now thought the Republicans had behaved unreasonably and irresponsibly.

"We accomplished that," a senior administration official said.

As Axelrod saw it—after poring over reams of data from polls and focus groups—achieving above-the-frayness was not the only way for Obama to impress independent voters. These voters wanted cooperation in Washington, but they also yearned for results. If Obama fought vigorously for a jobs bill containing a variety of popular provisions—many of them centrist in nature—his standing among independent voters would likely improve.

"The Republicans," Axelrod maintained, "have locked themselves into a bad economic theory that neither independent voters nor Democrats accept—that the trickle-down thesis is more fresh and new than the last decade or the 1920s." And focus groups of independent and Democratic voters especially liked the moment in the jobs speech when the president declared he would take the fight for the jobs bill "to every corner of this country."

Obama's poll numbers did not shoot up after the jobs speech and the subsequent barnstorming. White House aides claimed they were not worried—at least not too much. "We need good numbers in October 2012, not October 2011," a senior administration official said.

Plouffe was a practitioner of the long game, and he, Pfeiffer, and the rest of the political team had developed several objectives to enhance the president's standing with voters, particularly those fickle independent voters—and each one they believed was within Obama's reach.

The first aim was to bolster voters' faith that Obama was focused on jobs. "This was always a problem for us," a senior administration official conceded, especially during Obama's push for health care reform. Through the past year, Obama and his aides realized, the public had rarely seen Obama striving to create jobs. When he reached their living rooms, via the media and the Internet, it was because of uprisings in the Arab world, the bin Laden operation, or the near government shutdown and subsequent debt-ceiling dispute. If he now zeroed in on jobs, jobs, jobs, his advisers believed, he could indeed strengthen voters' impression that he fully shared their top concern.

Another goal was to persuade voters that Obama had an effective and ambitious plan and the Republicans did not—and that the other side was more interested in thwarting the president than in forging bipartisan compromises that would address the jobs deficit. The debt-ceiling dustup had indicated that Obama could register gains in public sentiment relative to the Republicans.

A third aim was to address, as one senior administration official put it, "the narrative that Obama was weak through the debt fight." White House aides believed that this impression was not widespread within the electorate. But they recognized it was a common gripe among elite Democratic, progressive, and media circles—and thus could spread to a wider audience. To try to beat back this sentiment, Obama and his aides were looking to throw some hard punches at the Republicans.

A major shift in voter perception of Obama would not happen immediately. This was a blueprint to address voter attitudes in a pragmatic and gradual fashion.

Obama's new tactics, not surprisingly, were not having much impact on Capitol Hill. In early October, Cantor said that the House Republicans would not afford the president a floor vote on his full package.

At a Texas rally, Obama excoriated the House majority leader: "I'd like Mr. Cantor to come down here to Dallas and explain what in this jobs bill does he not believe in. . . . Does he not believe in rebuilding America's roads and bridges? Does he not believe in tax breaks for small businesses, or efforts to help our veterans? . . . Tell small business owners and workers in this community why you'd rather defend tax breaks for folks who don't need them—for millionaires—rather than tax cuts for the middle-class families."

Two weeks later, Senate Republicans mounted a filibuster and blocked the American Jobs Act, which was backed by a narrow 51-vote majority.

That day, two senior White House officials held a background briefing for reporters and claimed it was "a very dangerous moment for the Republican Party." They pointed to polls that showed support for components of the jobs bill in the 60 percent to 80 percent range. Yet the Republicans still seemed intent on thwarting much of it—perhaps even the payroll tax cut.

The notion that the Republicans can "oppose everything and not suffer is wrong," one of the officials maintained. Listing the elements of the package—teachers, construction workers, unemployment insurance, rebuilding schools, tax cuts for small businesses—one of the aides noted, "We're going to have thirteen months of political debate: How much do we value these things?" Would voters next year "want to support a candidate who believes we should starve education, starve our infrastructure, and give huge tax cuts to millionaires paid for by senior citizens?"

OBAMA WAS BACK ON THE BUS. IN ASHEVILLE, NORTH CAROlina, he visited the regional airport and highlighted the need to renovate its sole runway. The $50 billion in the American Jobs Act for infrastructure included $2 billion for airports. At West Wilkes High School in Wilkesboro, North Carolina, Obama emphasized preventing teacher layoffs. The jobs bill provided funding to support nearly 400,000 educator jobs nationwide. North Carolina, he noted, would receive funds to cover 13,400 teachers and school workers across the state. And so on.

For months, Plouffe had been telling nervous Democrats that reelection could be won, even with an awful economy, if Obama's crew developed customized organizing, communications, and policy strategies for targeted swing states. This was the beginning.

AS OBAMA ADOPTED A MORE POPULIST MESSAGE, HE AND HIS political aides were watching the burgeoning Occupy Wall Street movement against corporate greed, an inchoate protest that had led to encampments and demonstrations in New York City and elsewhere. Obama publicly noted that he sympathized with the demonstrators' broad grievances covering income inequality, unemployment, and economic injustice. The White House said Obama was looking to protect "the interests of the 99 percent of Americans," acknowledging, if not adopting, the rallying cry of the Occupy Wall Streeters.

"We knew from focus groups and polling that people had never been more angry about inequality and Wall Street," a senior administration official said. "The sentiments behind Occupy Wall Street were what we hear all the time: the system is rigged, the rich get richer, Wall Street screwed up but it ended up okay. Some of that is directed at the president. But on a comparative basis with the Republicans, we do well on those issues."

Obama's foes on the right, though, were using the Occupy movement as a club against the president and his supposed promotion of class warfare. In what was perhaps the most outlandish of these efforts, the *Wall Street Journal* published an op-ed by David Moore, the CEO of a Wall Street holding company, who was appalled when a New York City panhandler refused his offer of a single dollar and shouted at him, "You Wall Street fat cats." Who was responsible for this great offense? The president, of course.

Moore angrily contended that "the president's incendiary message has now reached the streets" and that would "seriously undermine the chances for reasonable compromise."

The chances for reasonable compromise had evaporated during the debt-ceiling negotiations. And the Republicans were now not showing much interest in the components of Obama's jobs package. Senate GOPers blocked—via filibuster—the first breakaway legislation from the American Jobs Act, a bill to protect teachers and first responders from layoffs. Two weeks later, they did the same with legislation embodying the infrastructure measures in the jobs plan.

OBAMA COULD KEEP PRODDING THE REPUBLICANS TO PASS HIS jobs plan or parts of the package. Yet in the 24/7 world of the Internet and cable news, how often could the president swat the

Republicans before his demand, politically popular or not, would become stale and he might appear impotent? He needed a second act for his revival mission.

In a meeting with aides, Obama casually mentioned, "I'm not going to sit around and do nothing while they sit on their hands. The country can't wait. We can't wait."

He told his staff he wanted to start issuing as many executive orders as he could.

His advisers told him the policy changes he could accomplish with executive orders would be small compared with the scope of the economic troubles confronting the nation.

Obama argued back: "What I can say is that I'm pushing Congress to do this. But we can't wait, so we're doing these other things."

And he began a flurry of other things. In late October 2011, during trips to Las Vegas and Denver, Obama announced a series of We Can't Wait initiatives he was implementing by executive order. The first was a rules change that would help underwater homeowners refinance their mortgages. The administration's foreclosure programs, riddled with assorted problems and failing to reach many in-trouble homeowners, had been disappointing, even for Obama (who had acknowledged during a Twitter town hall the previous July that his administration had erred in not moving more vigorously to address the mortgage crisis). The new rules, though, could help up to 1.6 million homeowners.

The second action was a set of rules that would permit more than one million people who had borrowed money to pay for college to reduce their monthly payments and allow six million students and recent college graduates to consolidate loans and reduce interest rates.

In the weeks ahead, Obama would wield his executive power to direct community health centers to hire thousands of veterans, government agencies to curb wasteful spending (no unnecessary

BlackBerrys or agency swag!), and the Food and Drug Administration to reduce drug shortages. Under the We Can't Wait banner,
the White House would also announce stronger fuel economy
and greenhouse gas pollution standards for cars and trucks.

This whole campaign had a touch of hype to it. Obama was
issuing executive orders at about the same pace as the past two
occupants of the White House. None were world-changing. But
each order could affect thousands here or a million or so there—
and that would add up.

This continuous promotion of initiatives, a Democratic strategist who worked with the White House noted, "showed that
Obama was not just a captive of congressional dysfunction. . . .
If the Republicans are crazy and intractable, you're going to keep
working every day. You can't spend thirteen months saying you
have to pass my jobs package."

In terms of winning over voters—that is, winning them
back—Obama was seeking progress inch by inch. Fighting for
the payroll tax cut for the benefit of 160 million Americans, decrying a do-nothing Congress, issuing executive orders to help
homeowners, students, and veterans—the political benefit would
be cumulative.

In early November, a *Washington Post*–ABC News poll suggested that this strategy could pay off. Half the people surveyed
agreed with the proposition that Obama had been "making a
good faith effort to deal with the country's economic problems,
but the Republicans in Congress are playing politics by blocking
his proposals and programs." Forty-four percent cut the other
way: Obama had not "provided leadership on the economy, and
he is just blaming the Republicans in Congress as an excuse for
not doing his job."

But 54 percent of independents had a favorable impression of
Obama's effort, compared with 40 percent who did not.

This was incremental change Plouffe could believe in.

IN MID-NOVEMBER, THE REPUBLICANS JOINED DEMOCRATS IN voting for a small slice of Obama's jobs package—legislation that offered tax credits to companies that hire unemployed veterans and repealed a tax on government contractors. But this represented less than 0.5 percent of the total jobs bill.

Obama and his aides believed the GOPers would have to yield on the payroll tax cut before Congress left town at the end of the year, and the president kept pounding the Republicans for not supporting this part of the package. But the Tea Party–dominated Republicans were not coming around.

Some GOP lawmakers were searching for a substitute for the millionaires' surtax or considering what payoff they could extract in return for passage of this tax cut; other GOPers simply didn't want the extension at all. They remained stuck in a world where cutting spending, bulldozing government, and protecting the rich from higher taxes were the avenues to economic expansion.

IN LATE NOVEMBER, THE INEVITABLE HAPPENED: THE SUPER-committee failed. Hours before the midnight deadline, the Democratic and Republican cochairs of the supercommittee released a statement saying they were "deeply disappointed" the committee could not produce an agreement. If nothing intervened, in January 2013 the trigger would go off: $1.2 trillion in automatic cuts, half within the Pentagon.

The supercommittee had been doomed from the start. Before its members could put on their green eyeshades, Boehner had declared that all deficit reduction had to come from slashing federal agency spending and shrinking entitlements—no revenues.

Obama and his aides had reasonably concluded that the Republicans would never accept significant revenues—and who would know better?—and presumed the supercommittee would end a flop. So the president had kept his distance—and had deftly avoided being ensnared in yet another unwinnable debate over deficit reduction and becoming tainted again by Washington failure.

The day the supercommittee crashed, Obama hit the briefing room. He did not blame a dysfunctional Congress, as many commentators rushed to do—or as he might have done earlier in the year. Instead, he pointed an accusatory finger at congressional GOPers: "There's still too many Republicans in Congress who have refused to listen to the voices of reason and compromise" and who continue to protect "tax cuts for the wealthiest 2 percent of Americans at any cost, even if it means reducing the deficit with deep cuts to things like education and medical research. Even if it means deep cuts in Medicare."

And he issued a threat: "I will veto any effort to get rid of those automatic spending cuts to domestic and defense spending. There will be no easy off-ramps on this one."

The Republicans predictably blamed Obama for somehow not leading the supercommittee to the promised land—even though Republicans on the panel had privately requested he stay away. Obama had obviously made the right call, and this moment undercut what had become a familiar complaint: that Obama was a terrible negotiator.

Earlier in the year, the president had been unable to escape the debt-ceiling trap that reaffirmed the GOP mantra that deficit reduction was the most pressing economic priority of the day. But by shrewdly negotiating a trigger with more than half a trillion dollars in military spending cuts, Obama and his aides might have pulled a fast one on the Republicans—just as they had done during the tussle over the 2011 spending levels.

Now, hawkish Republicans would be looking to escape the deal. And with his vow to veto any attempt to defuse the automatic cuts, Obama was becoming Washington's fiscal enforcer, who would not let the Republicans squirm their way out of their commitment to blot up Washington's red ink. This would be a major turnaround.

Obama could expect to face protests within the military about this across-the-board slash. Leon Panetta, the new defense secretary, was already hollering about the potential impact of the automatic cuts. And Pentagon-protecting Republicans, no doubt, would try to undo or remake the trigger—and warm up talking points casting the president as a weak-on-defense Democrat. But having dispatched Osama bin Laden and other al-Qaeda leaders, as well as launched the military action that ended Qaddafi's regime, the president was in a good position to beat back this traditional Republican assault. Moreover, Obama could declare, "A deal's a deal"—and offer the Republicans a way out only if they would raise revenues to cover the military funds they wished to defend.

In the deficit-reduction wars, Obama had what his aides considered another advantage: the expiration of the Bush tax cuts. They were scheduled to disappear at the end of 2012. If Obama did nothing, he could reduce the deficit without touching Medicare, Medicaid, or Social Security benefits. The end of the Bush tax cuts along with the trigger (presuming it was not uncocked) would produce a flood of deficit reduction.

Republicans might again try to hold the extension of middle-class tax cuts hostage to preserve the tax breaks for wealthy Americans. But Obama was determined not to be held captive next time. He had declared in private and public that the tax-cut compromise of December 2010 was a onetime shot. And he was now telling his aides that he would veto any legislation sent to him—before or after the 2012 election—that continued the Bush top-bracket tax cuts.

"He will send it back to the Republicans until they break," a senior administration official remarked.

The deficit wars that defined Obama's third year in office were far from over. The biggest disputes had only been delayed. In early 2013, the three most explosive clashes would converge. The Bush tax cuts would die, the automatic cuts would kick in (threatening the Pentagon), and the nation's debt would be reaching the limit set in the August deal. It could be a storm of fiscal and political commotion. Obama would confront all this after the November election, either as a lame-duck president or as a reelected chief executive.

It was nothing for him—or anyone else—to look forward to.

For several weeks, Obama had been thinking of updating a speech he delivered in 2007 as a presidential candidate at Nasdaq's corporate headquarters in New York City.

In front of about 150 Wall Street executives, Obama had asserted that corporate excesses (such as extravagant CEO compensation and the subprime mortgage fiasco) undermine American capitalism. He told the financiers that in the modern economy the fortunes of Wall Street and Main Street were interconnected. He called for a "re-appraisal of our values as a nation." This was a prescient address, with Obama warning that the subprime debacle and similar Wall Street shenanigans could set off a larger crisis—which was precisely what happened a year later.

Perhaps the Occupy Wall Street movement had caused Obama to consider reprising this speech. And a Congressional Budget Office report on income inequality years in the making had caught his eye. It noted that the income of the wealthiest 1 percent in the United States had nearly tripled since 1979. In that period, the after-tax income of middle-class Americans rose only about 1 percent a year. The rich were boosting their annual take at about seven times the rate of average American families.

For years, Obama had been thinking and speechifying about the loss of economic security for the middle class and the apparent decline of social mobility. Reviving the American Dream for middle-class families had been one theme of his presidential campaign. The president now wanted to return to it.

On December 6, 2011, Obama flew to Osawatomie, Kansas, a small town (population 4,447) with two claims to history. In 1856, abolitionist John Brown and a few dozen comrades mounted a losing fight there against a 250-man pro-slavery militia that attacked the town. And on August 31, 1910, as part of a two-day commemoration of John Brown, President Teddy Roosevelt, then out of office but still deeply involved in politics, delivered a speech there defining a "New Nationalism."

Roosevelt's remarks were a bold declaration of progressivism at a time when the country was still undergoing a profound transformation from agrarian society to industrial powerhouse. Before a crowd estimated to be about thirty thousand, T.R. denounced corporate "special interests," declared that the "citizens of the United States must effectively control the mighty commercial forces," and called for a "square deal" for workers, which would include rigorous government regulation of the workplace and Big Finance.

Roosevelt's address was probably the most radical ever delivered by an American president in or out of office. It helped define early twentieth-century liberalism. After his speech in Osawatomie—assailed at the time as "communistic," "socialistic," and "anarchistic"—Roosevelt broke with the Republican Party, formed the Progressive Party, and ran against incumbent Republican president William Howard Taft in the 1912 presidential contest, which Woodrow Wilson, a Democrat, would win.

Obama had long admired Roosevelt's "New Nationalism"

speech. And in the midst of his clash with Republicans over the payroll tax cut—the White House had posted on its website a countdown clock marking the time left before this tax break for millions of Americans would expire—the president sought to place the battles of the moment into a larger context.

At Osawatomie High School, Obama noted that the recent economic collapse had precipitated "a raging debate over the best way to restore growth and prosperity, restore balance, restore fairness." He said this was "a make-or-break moment for the middle class and for all of those who are fighting to get into the middle class."

On one side of this debate, Obama asserted, was a "certain crowd in Washington" that was peddling a familiar refrain: " 'The market will take care of everything,' they tell us. If we could just cut more regulations and cut more taxes—especially for the wealthy—our economy will grow stronger. . . . Jobs and prosperity will eventually trickle down to everybody else. And, they argue, if prosperity doesn't trickle down, well, that's the price of liberty."

He acknowledged that this "simple theory" speaks to the tough individualism and skepticism toward government that can be found in "America's DNA." But this theory, he warned, failed the nation in the 1920s and the 2000s.

The United States, Obama pointed out, had become a nation of greater economic inequality. A child born into poverty today had only a one-in-three chance of reaching the middle class. After World War II, such a child had a slightly greater than fifty-fifty shot.

Obama contended that the Republicans' let-the-market-rule view would not restore "middle-class security." He was back to the vision he had presented in his State of the Union speech eleven months earlier: to revive a strong and prosperous middle class, Americans must join together—through government—to educate, innovate, and build. This would require money, and that meant more taxes on those who could bear the burden.

"This isn't about class warfare," he said. "This is about the nation's welfare."

Obama, whose mother's family had come from Kansas, was not rabble-rousing, as Roosevelt had done 101 years earlier. His speech hardly matched Roosevelt's in reach or fierceness. But Obama decried banks that were fighting the new safeguards his administration had imposed on banks and securities firms. He blasted Republicans for blocking Richard Cordray's nomination to head the Consumer Financial Protection Bureau. (The GOPers didn't object to Cordray; they despised the CFPB. Weeks later, Obama would place Cordray in the post with a recess appointment that enraged Republicans.) He vowed to back legislation that would create tougher penalties for Wall Streeters caught violating fraud laws.

Most of all, the president called for a vision based on communal advance, not unfettered business, noting this was fundamentally a matter of fairness—a message Obama and his aides believed would play well in the coming year with both the Democratic base and the political middle.

"In the end," he said, "rebuilding this economy based on fair play, a fair shot, and a fair share will require all of us to see that we have a stake in each other's success." That, he added, was a matter of "American values."

Obama's six-thousand-word speech, which received only moderate media attention, would not change the immediate political dynamic. Speeches tend not to do so. "Words count for nothing except in so far as they represent acts," Roosevelt had said in Osawatomie. And this address did not boldly offer new ideas. Obama was not redefining his presidency, reshaping progressivism, or coining a slogan as marketable as "New Nationalism." But it did provide a full picture of what Obama was thinking.

At the start of his third year in office, the president in his State of the Union speech had declared his vision of how best to prepare the economy for the long run, and throughout the budget and debt-ceiling battles, Obama had often tethered his stands to progressive values. His sharp-edged anti-Ryan speech was a values-based counter to the Republicans' crusade for spending cuts and government downsizing. But for much of this turbulent year, Obama had not often had a clear shot at voicing his story of America.

This address marked a full integration of his vision and values. Afterward, Jim Messina, Obama's reelection campaign chief, sent out an e-mail to Obama supporters proclaiming that the approach Obama presented in this speech "will inform every discussion we have with undecided voters over the next year."

"WHY NOT DO THE RIGHT THING FOR THE AMERICAN PEOPLE even though it's not exactly what we want?" Boehner said, three days before Christmas. With that ill-advised remark, the speaker surrendered in the payroll tax cut battle Obama had initiated—at least, in the first round.

For weeks, the president had been pounding the Republicans for refusing to vote for legislation that would continue the payroll tax cut and extend unemployment benefits. He repeatedly asked whether the Republicans, who always fought bitterly to protect tax breaks for the wealthy, would let taxes rise on middle- and low-income Americans.

For some Republicans, the answer was yes. Tea Partiers in Boehner's caucus (and several Senate Republicans) openly said they did not support the payroll tax cut extension—in essence arguing that tax breaks for working-class Americans, as opposed to tax cuts for the rich, do not stimulate the economy. GOPers also did not fancy applying a tax on millionaires to pay for it. In one of the year's weirder votes, more than half of the Senate Republicans in early December voted against a Republican-proposed bill to continue the payroll tax reduction.

On the House side, Boehner was in another tight corner. He said the payroll tax cut was worthwhile, but his caucus was once again bucking him. So he did what he usually did. He tried to put the squeeze on the president.

Boehner assembled a bill that would extend the payroll tax cut and unemployment benefits for a year and fix a Medicare wrinkle that would lead to automatic cuts to doctors—but only if the Senate and the president accepted a conservative wish list, including various spending cuts (some targeting Obama's health care overhaul), riders to nullify certain environmental regulations, a boost in Medicare premiums for middle-class beneficiaries, and a provision that would force the administration to decide within sixty days whether to green-light the controversial Keystone XL pipeline (which would run from Canada through the Plains states to the Gulf of Mexico).

Boehner managed to rally his House GOPers to pass this bill on December 13—ignoring Obama's veto threat and Harry Reid's warning that the bill would be DOA in the Senate.

With time running short before the Christmas break—and the expiration of the payroll tax cut and unemployment insurance benefits—the White House blasted the Republicans for playing political games. Simultaneously, another squabble over a government spending bill threatened to shut down the federal government during the holiday season. It was a fitting end to the year: more chaos and conflict, much of it driven by the extremist wing of Boehner's party.

The House and Senate finally managed to vote for an omnibus spending bill funding the government through the following September. At the same time, Harry Reid and Mitch McConnell, realizing they couldn't forge a full agreement on the payroll tax matter before the holidays, crafted a compromise to extend this tax cut and unemployment benefits for two months.

Their bill dumped the surtax on millionaires—instead the cost of the tax cut would be covered by boosting fees lenders pay to Freddie Mac and Fannie Mae—and it included the Keystone XL pipeline measure. They presumed they'd later find a way to work out something for the rest of the year. On December 17, the measure passed 89–10.

Weeks earlier, Obama had said he would oppose an effort to tie the pipeline decision to the payroll tax cut, but he didn't complain about the pipeline provision, which didn't actually force him to approve the project. The Republicans might have even set back the project, for the State Department announced that if it were compelled to issue a decision within sixty days, it would have to withhold approval, given that there would be insufficient time to conduct a review of the pipeline. (A month later, that's what the State Department would do.)

Obama embraced the temporary extension, figuring that he had won the fundamental debate—whether or not to extend this tax cut in 2012—and that it would be rather difficult for the Republicans to not go all the way when they returned the next year. He hailed the measure as a rousing bipartisan success story—thirty-nine Senate Republicans had endorsed it—and pressured Boehner, noting he expected House Republicans to sign off on it right away.

Boehner seemed to expect that too, at first, and he called the short-term extension a "good deal" and a "victory."

Then during a private conference call with Boehner, many rank-and-file House Republicans expressed anger at Senate Republicans for ducking a fight. This compromise was missing many of the sweeteners Boehner had earlier used to win over his Tea Partiers. Moreover, a temporary continuation would give Obama more time to bash them for not getting this job done.

They let Boehner know that if he pushed them to vote on the Senate bipartisan compromise, there would be an uprising.

The final showdown of the year was at hand. Boehner's Republicans were in a standoff against their fellow Republican, Mitch McConnell, who was allied with Reid and Obama. And Boehner was isolated—at the mercy of his Jacobins.

The speaker quickly retreated from his support of the Senate compromise (that he actually had empowered McConnell to produce). He and other House Republicans mounted a campaign against the deal, claiming this temporary extension was a gimmick and undermined certainty.

They contended that the Senate and the president should join them in composing a yearlong solution, as they had done weeks ago on their own. (Their arguments were a bit disingenuous; Boehner and other Republicans had earlier derided the payroll tax cut as bad policy and the previous summer had advocated a short-term hike in the debt ceiling.)

If no extension were passed and middle-class Americans saw their tax bill go up on January 1, Boehner and his crew maintained, Obama and the Senate would be to blame.

This was another game of chicken. But Obama and Reid wouldn't blink. The Senate had essentially ended its session, and Reid refused to call it back to negotiate with the House Republicans. And Obama, sensing an opportunity to slam the House Republicans (whose approval rating had yet to recover from the debt-ceiling debacle), went on the offense.

The White House arranged for administration officials to blast Boehner and the Republicans in interviews with regional, African American, and Hispanic media. It encouraged Americans to send in stories to say what $40 a paycheck—the average amount of the payroll tax cut—would mean to them. (The White House at one point was receiving two thousand responses an hour, many of them heartfelt.) Obama believed the GOP had wildly miscalculated by initially dismissing the significance of what could be $160 a month extra for a dual-income family.

He assailed the House Republicans for not even permitting a vote on the bipartisan Senate measure. In a phone call to Boehner, Obama insisted, as the White House put it, that the "bipartisan compromise passed by almost the entire Senate is the only option to ensure that middle-class families aren't hit with a tax hike in ten days."

Obama was again a champion of bipartisan reasonableness. And this was an easy message skirmish for the Democrats to win: the House Republicans were willing to screw over middle-class taxpayers if the Senate and the president didn't accede to their extreme demands. Though prior GOP hostage taking had succeeded, this time Obama didn't have to fear economic catastrophe as a possible result *and* he had a divided Republican Party to exploit.

As the Tea Party–whipped Boehner stood with his comrades and arranged a procedural vote that rejected the Senate bill, he came under fierce attack—from other Republicans. John McCain huffed that the House GOP antics were "harming the Republican Party." Republican Senator Scott Brown remarked, "It angers me that House Republicans would rather continue playing politics than find solutions."

The conservative editorialists of the *Wall Street Journal* chastised Boehner and his team: "We wonder if they might end up re-electing the President before the 2012 campaign even begins in earnest." Karl Rove, Newt Gingrich, and other Republicans piled on. The GOP establishment was practically in mutiny against Boehner.

On December 22, Obama called Boehner and told him the only way out was for the House to pass the Senate bill. And

McConnell gave the speaker a soft kick, issuing a statement that called on the House to pass the temporary extension. But that morning Boehner's office insisted the speaker was not interested.

In the early afternoon, Obama, surrounded by Americans who had shared their $40 stories, said he was in full agreement with McConnell. Boehner and the House GOPers were the outliers. "The only reason" this tax cut was in jeopardy, the president maintained, was "because a faction of House Republicans have refused to support this compromise."

As Obama was confronting the Tea Party, Boehner was preparing to cave. He notified Senate leaders that he would bring a temporary extension to the House floor with a few minor technical differences from the Senate bill.

During a conference call with his fellow House Republicans—most of whom had left town for the holidays—Boehner explained that the bill would be brought up under a unanimous consent rule the next day. This meant it would pass without a roll-call vote, unless any member objected. But there was not enough time for many of his Tea Partiers to scramble back to DC.

After being embarrassed, Boehner was finally saying no to the Tea Party. During the conference call, the speaker kept the mute function on. His troops could not talk back to him.

At a wave-the-white-flag press conference later in the day, Boehner remarked, "Sometimes it is politically difficult to do the right thing."

Obama and his aides—in private—took credit for having trapped Boehner, citing three earlier White House decisions. The previous December, Obama, following Sperling's suggestion, had made the yearlong payroll tax holiday a central component of the lame-duck tax-cut compromise, realizing it would be tough for the Republicans to resist extending it.

Then, in the summer, Obama had held the line against any short-term lift for the debt ceiling, preventing a rerun of that spectacle from consuming all the political oxygen at the end of the year. This created space for a political scuffle that was to Obama's advantage. Finally, Obama's decision to propose and then relentlessly campaign for a large jobs package had placed pressure on the Republicans. In essence, he had provided the House Republicans the opportunity to display once more their rigidity—and they seized it.

The payroll tax cut battle was not over. It would start again in the new year. But as 2011 was ending, Obama had benefited from a profound Republican blunder that he had helped orchestrate.

The next day, Obama came to the briefing room and spoke for five minutes. It was a subtle victory lap. He called on Congress to work—"without drama"—on extending the tax cut and unemployment benefits for the full year.

"This continues to be a make-or-break moment for the middle class," he somberly noted, adding, "There are going to be some important debates next year, some tough fights. . . . But that's the kind of country that I'm fighting for—one where everybody has a fair chance and everybody is doing their fair share."

Boehner and the Tea Party had just handed him another chance to show that.

"Aloha," he said at the end.

He then left to join his family in Hawaii. He'd have a week off before beginning the final year of his first—and perhaps only—term.

Chapter Fourteen

THE BATTLE AHEAD

PRESIDENT OBAMA ENTERED 2011 HOPING TO MOVE QUICKLY to a values debate with the Republicans. But external circumstances—and Republican obstructionism—intervened.

After taking that shellacking in the midterm elections, the president and his aides freely admitted that they had failed to define their own achievements during their first two years and that they had not shaped the nation's political story. For much of the third year, they had not done better.

"We felt like we were so immersed in crisis in the first two years and were so much a triage unit, we didn't spend enough time presenting an overarching narrative about where he's going," David Axelrod recalled at the end of 2011.

"And we were muted in our communications for the first eight months of this year," he continued. "It's been much better since the jobs speech. In the first two years, we tried to communicate too many things. In the first eight months of this year, we tried to communicate too little. There was a lot of hunkering down with fiscal issues and there was a desire not to get into gratuitous back-and-forth [with the Republicans]."

With the American Jobs Act and Obama's subsequent barnstorming, the president was fully in the fray, not above it. He had finally reached the political spot he had anticipated a year earlier—though it had taken longer to get there and the pathway had been even bumpier than foreseen.

There had not been many policy successes, but the president had emerged with a clear strategy that included championing a progressive set of values and pressing a distinct vision of how to guide the nation and its economy into the twenty-first century.

Obama's Gallup presidential approval rating had remained stuck in the low 40s since the debt-ceiling fight, but he scored better in other polls than the Republicans when respondents were asked whom they trusted to handle the economy and protect the middle class. (A *Washington Post*–ABC News poll in mid-December 2011 found Obama's approval rating to be 49 percent, a nine-month high, excluding the post–bin Laden raid bump.)

And David Plouffe could barely believe that at the end of the year Obama's approach to deficit reduction was better received by the public than the Republicans'. He told people it was as if the Republicans had gained ground on the Democrats on the issue of children's health care. A core Republican strength had been neutralized.

OBAMA AND HIS AIDES CONSIDERED A STEADY AND SHARP DEPICtion of his vision and values the key to reelection. Though the unemployment rate in early December had dropped half a point to 8.6 percent (partly because droves of workers dropped out of the labor force), the economy was not likely to improve tremendously in 2012, and not since 1936, when Franklin Roosevelt won his second term, had a president been reelected with unemployment above 7.2 percent.

But his team had a plan.

A top campaign strategist for Obama described its three parts. First, the reelection effort would establish a proper context in which voters could review and understand Obama's first term.

"We have to remind people what Obama faced when he took office," this strategist said. "The economy was losing eight hundred thousand jobs a month. Where we are now is not exciting. But it's much better. We'll never convince people the world would be so much worse without him, but we can point to health care, Wall Street reform, and other accomplishments, and make sure our activists and volunteers have all these talking points so they can answer those people who say they are disappointed."

In a speech at the Economic Club of Washington, Gene Sperling made a similar point: "It may be that, quote, 'Prevented a Second Great Depression' does not read well on a bumper sticker. But it is an appropriate description of policy choices that dramatically improved and helped the lives of tens of millions of our fellow working families and the global economy."

Or as Axelrod quipped in one interview, "You know, his slogan should be, GM Is Alive and Bin Laden Is Dead."

The second step was to hammer that values and vision contrast with the Republican candidate. Perhaps Obama's most crucial reelection goal was to prevent the contest from becoming a referendum on him. The Republicans could easily argue that the economy was awful, and Obama should be booted. If unemployment remained high and economic growth continued to lag, this case for change would be compelling.

Obama had to shift the question to which candidate offered a better vision for the future and a more compelling declaration of the nation's values.

Throughout the fall of 2011, Obama had been comparing his vision and values with those of the Republicans. He was for the prudent use of government resources to produce jobs, enhance middle-class security, and make investments necessary for future economic progress (while tending to fiscal matters in a balanced manner). The Republicans favored strangling government, allowing Wall Street to again run free, and protecting tax breaks for the wealthy.

The debt-ceiling fight demonstrated that Obama could best the Republicans on these fronts. And in early December 2011, Plouffe was delighted when Mitt Romney enthusiastically renewed his endorsement of Paul Ryan's budget. This embodied the values contrast Obama was looking to establish.

The third step was fighting. For much of 2011, Obama was bogged down in negotiations he and his aides believed were crucial to avoiding economic disaster. To win reelection, he'd have to persuade voters he had the mettle and determination to battle for them.

Independent voters, as Obama's aides saw it, desired compromise and an end to partisan sniping—but only if decent results were realized.

"They do not want capitulation to the Tea Party in the name of compromise," the Obama strategist remarked. "These people prefer a postpartisan world, but in the face of just-say-no Republicans, they want someone who will be very direct."

In his first three years in office, Obama had taken a lot of guff for not being a fighter, for being too quick to conciliate and seek an accommodation.

"A big part of the game is keeping our people engaged and enthused, and you can't do that if they don't think you're fighting for them," a senior House Democratic aide complained about Obama during the debt-ceiling brawl. "They don't respond to the pragmatic management of government. You don't always get rewarded in politics for being rational."

But Obama could be a steely battler—though in a cool and quiet manner. During the Egyptian and Libyan crises, after careful deliberation, he took decisive steps. He did the same when he launched the bin Laden mission that could have ended his presidency had it failed. Yet following these episodes, he displayed little swagger, and the full extent of his behind-the-scenes fortitude did not completely register with the public. But after Romney accused Obama of "appeasement," the president did snap back, "Ask Osama bin Laden and the twenty-two out of thirty top al-Qaeda leaders who've been taken off the field whether I engage in appeasement."

Obama spent much of his presidency seeking consensus perforce. But he was a competitor. The 2008 campaign had not been merely an upbeat and touchy-feely parade of hope-iness, yes-we-can-ism. Obama fiercely took on establishment favorite Hillary Clinton and then war hero John McCain, and his campaign machine was always aggressive.

Obama's advisers knew he could be a fighter. The Republican presidential nominee would definitely try to make the election a referendum on Obama; the president and his allies would have to promote his contrasting vision *and* tear into the Republican nominee. Obama could do that.

In Chicago, Jim Messina, who left the White House shortly after the repeal of Don't Ask/Don't Tell, was in charge of the reelection campaign. In preparation, he had read nearly every book he could on presidential campaigns of the past hundred years. He had found great lessons in the reelections of Franklin Roosevelt, Bill Clinton, and George W. Bush—but he would not share these with reporters who asked. He had, however, reached a traditional conclusion: winning campaigns are always about the future.

Messina believed—and polls backed this up—that a majority of voters shared Obama's vision and values, as evidenced in the debt-ceiling showdown and the ongoing push for the jobs package. The election in 2012, he thought, need not be an up-or-down vote on the incumbent, as long as the Obama campaign robustly conveyed its core messages. Messina even thought Obama could be the hope and change candidate in the race—though, of course, with a modification in expectations.

The hope and change sales pitch in 2008 had allowed Obama to define himself in assorted ways—and, more important, permitted voters to embrace him for different reasons. Now, the president had a concrete record that could be judged. If Messina and his crew were serious about reprising the change theme, they had to convince voters the president's actions had truly resulted in positive change.

Saving the economy from depression, overhauling health care, reforming Wall Street, enacting a stimulus that created or saved millions of jobs—none of that had helped the Democrats in 2010. But Messina was planning an aggressive strategy to trumpet Obama's accomplishments. Even on health care: Do you want Republicans to take away your mammograms and colonoscopies or your insurance for preexisting conditions?

Obama could still hold on to the mantle of change, Messina insisted, by persuading voters that the government needs to do big things to create jobs, develop clean energy technology, confront immigration reform, invest in education. With such an approach, Obama's aides contended, the president could both court independents and rev up the Democratic and progressive base.

There were obvious obstacles. Though White House aides routinely cited polls showing that the president's Democratic base supported him at high levels, these surveys also revealed a so-called enthusiasm gap, with Democratic voters expressing less passion about the 2012 contest than Republican voters. And a number of Democratic and progressive activists and influentials were disappointed with the president and the outcomes of the first three years. "We'd be out of our mind to say this is not real," Messina said.

By the end of 2011, the campaign had already developed mechanisms and metrics to respond to the disillusionment. As part of its grassroots operation, the reelection effort was conducting one-on-one conversations between Obama volunteers and prospective voters—with the results of the conversations sent back to the Chicago headquarters and stored in a database that logged the concerns, pet issues, and past voting histories of each voter. By mid-November, the campaign had amassed one million of these contacts, and many of these conversations had addressed voters' discouragement concerning the president.

The leading causes of the disenchantment tended to be Obama's abandonment of the public option and the lack of progress on climate change, immigration reform, and Guantánamo. (Obviously, these voters were liberal Democrats.) With all this data in hand, the campaign was developing strategies for its volunteers to address these concerns. Surely, once the Republican nominee emerged and the alternative was clear—a Republican in the White House with possibly the Tea Partied GOP controlling both sides of Capitol Hill—Democratic enthusiasm would increase. But Messina and the campaign were not willing to wait.

There were positive signs for the campaign. In the first ten months of 2011, a million people donated to the campaign, with 46 percent having never before contributed to Obama. By the end of the year, the campaign had established on-the-ground operations in all fifty states and deployed more than a thousand neighborhood team leaders, attracting more volunteers than it had initially anticipated. And in November, Obama for America volunteers in North Carolina quietly mounted an organizing effort across the state and helped Democrats trounce Republicans in local elections. It was a test of the reelection campaign's theory that targeted grassroots activity in key areas could help win the day. There was little doubt that Obama's team could pull together a monster of a campaign organization.

STILL, OBAMA AND THE CAMPAIGN WOULD BE IN UNCHARTED waters, striving to reelect a president at a time of economic misery and profound popular doubt. (At the start of 2012, about 70 percent of Americans worried that the country was on the wrong track.) The combination of clever and sophisticated electoral strategies, cutting-edge tactics, a highly organized on-the-ground army, a bountiful campaign treasury, and an appealing message of vision and values would be up against a powerful impulse within the electorate to make a change . . . at the top.

In Obama's third year as the forty-fourth president of the United States, he did not triumph over the Republicans. Though he often outmaneuvered them—including during the payroll tax cut dustup at the end of the year—the GOPers, driven by unrelenting Tea Party extremism, had defined the debate for much of this period.

Yet as 2012 began, control of that debate was up for grabs—and Obama had reached the point where he had the chance to bolster his recently acquired advantage over the Republicans. Most of the public believed he had behaved more reasonably and responsibly during the absurd debt-ceiling trauma. (In mid-December, Pfeiffer tweeted, "Just a thought: if the House GOP had prevailed in August, right now we wld be debating the debt ceiling and possible default. Happy Holidays.") And Obama finally had managed to launch that war of ideas.

Looking at the 2012 election, Plouffe saw it evolving as a contest of greater clarity and contrast than the 2008 campaign. That heartened him. The more contrast, the better for Obama—and the millions of Americans who were with him.

The president's challenge in the final year of his first term was not merely to draw this fundamental comparison, but to persuade Americans that the survival of the middle class and the future of the nation was indeed at stake and that this face-off between him and the Republicans was not merely the same-old same-old partisan showdown or an interesting rivalry of progressive and conservative perspectives. And in his feisty, populist-tinged, and policy-drenched State of the Union address in January 2012, Obama aimed to establish this thread: "We can either settle for a country where a shrinking number of people do really well while a growing number of Americans barely get by, or we can restore an economy where everyone gets a fair shot." He noted there was a choice. Government could help the top 2 percent by preserving tax breaks for the wealthy, or it could bolster the middle class by financing programs that boost manufacturing, education, job-training, innovation, and research and that preserve the social safety net. "We can't do both," he said. What's at stake, he contended, is "American values."

At a time of profound trial for the United States, there were two distinct pathways forward. That's the story Obama would have to tell and sell—and the voters would get to choose.

"OBAMA RAN FOR PRESIDENT BECAUSE HE BELIEVED THERE WAS a confluence of problems that were a long time in the making, a consequence of rapid changes in communications, technology, and the economy," Axelrod said. "And the real question was, Are we mature enough as a country to deal with that in a way that works for most Americans?"

In his first three years in office, Axelrod insisted, Obama had been "exactly what he said he would be." He put results ahead of politics, outcomes above theatrics. He resisted the temptation to score cheap political points. He cut deals to solve problems—and to prevent harm to millions of Americans.

Axelrod summed up the case for Obama: "This may not be entirely satisfying, but he believes his highest responsibility is to get things done." And he noted that his own job, as Obama's message guru in 2012, was to "find a way to convey this, to politically monetize character." He added, "It's not entirely apparent you can do that."

Though the year 2011, full of political and policy complications and disappointments, had not always proceeded as anticipated, Obama had eventually succeeded in laying the foundation for the decisive campaign to come. In the face of unremitting opposition—and perhaps because of it—Obama had realized much of his strategic plan.

He could only hope the same would happen in 2012.

EPILOGUE TO THE
PAPERBACK EDITION

KID ROCK'S ANTHEM "BORN FREE" WAS BLASTING, AND MITT Romney took the stage at the Radisson Hotel in Manchester, New Hampshire. It was late April, and the former governor of Massachusetts had essentially wrapped up the Republican presidential nomination with several primary wins that night. He was still short of the total delegates needed, but his last serious rival, former Senator Rick Santorum, had dropped out two weeks earlier. Romney was clearly President Obama's opposition in November. And now in a victory speech geared toward the general election ahead, he was agreeing with Obama that the contest presented voters a fundamental choice.

Standing in front of a sign that said "A Better America Begins Tonight," Romney declared, "We know that this election is about the kind of America we will live in and the kind of America we will leave to future generations. When it comes to the character of America, President Obama and I have very different visions."

Obama had spent the previous year and a half setting up the 2012 campaign as a contest of visions, and here was Romney reinforcing the notion. Of course, Romney then dismissed Obama's vision by depicting it in scare-mongering terms:

> Government is at the center of his vision. It dispenses
> the benefits, borrows what it cannot take, and con-
> sumes a greater and greater share of the economy. With
> Obamacare fully installed, government will come to
> control half the economy, and we will have effectively
> ceased to be a free enterprise society.

This factoid had been cooked up with the phoniest of math
by Romney's campaign. Bruce Bartlett, who had served as a
senior economist in the Ronald Reagan and George H.W. Bush
administrations, would later note, "This analysis is so stupid it is
hard to know where to begin."

But accuracy was not Romney's goal. He wanted people to
believe the president intended to transform the United States
into a nation "where our lives will be ruled by bureaucrats and
boards, commissions and czars." He contended that "those who
promise to spread the wealth around only ever succeed in spreading poverty"—neglecting to mention that Medicare and Social
Security had done much to reduce poverty in the United States.

Romney was presenting a caricature of Obama's vision, while
offering tropes regarding his own: "I have a very different vision
for America, and of our future. It is an America driven by freedom,
where free people, pursuing happiness in their own unique ways,
create free enterprises that employ more and more Americans."

In the speech, Romney said the president should not be reelected because he had not succeeded in fully turning around the
nation's economic fortunes. But he also engaged Obama on the
values front—which could be beneficial for the president, who
wanted the election to be seen not as a referendum on his performance as the economy's caretaker-in-chief but as a face-off
between two candidates with differing ideas about basic policies.

Throughout 2011, Obama and his team had put great effort
into getting voters to perceive a profound conflict in values

between Obama and his political opposition. Romney was bolstering that point.

"CHICAGO IS STARTING TO WORRY," AN ADMINISTRATION OFFIcial had said several weeks earlier, referring to Obama's reelection headquarters.

At the time, Romney seemed wounded. The wild-and-wooly (and occasionally wacky) GOP primary contest had been tough on Romney and had compelled the onetime moderate Massachusetts governor to pander to the Tea Party-ized Republican base. He had been pounded for months for his long list of left-to-right flip-flops (gay rights, abortion, gun control, climate change). He was assailed for having presided over Bain Capital, a private equity firm decried as the epitome of vulture capitalism, and denounced for refusing to disclose years of his income tax returns. He was relentlessly slammed for implementing an Obamacare-like health care initiative in Massachusetts. His GOP rivals and critics on the right derided him as a faux conservative and a say-anything, inauthentic pol. On the campaign trail, he frequently came across as awkward and detached—a one percenter and a caricature of a protect-the-rich Republican politician. In a weak field, he had stumbled across the finish line bruised and bloodied.

Yet Obama, Jim Messina, David Axelrod, David Plouffe, and other top Obama strategists knew that a lot of voters had not paid attention to the GOP race and were open to a new guy. In March, Eric Fehrnstrom, Romney's top message adviser, had declared that once the primary battle was done, "everything changes. It's almost like an Etch A Sketch. You can kind of shake it up and restart all over again." Fehrnstrom was widely panned for the remark, which suggested that Romney was as sturdy as a column of sand.

In a way Fehrnstrom was right. Romney could reset once

he dispatched his pesky opponents. Obama's advisers believed that spring and early summer were the time for the Romney campaign to relaunch its product. The sniping from the right would be over, and with the gridlocked Congress doing little in Washington—in February the House Republicans had caved and agreed to a full extension of Obama's payroll tax cut—there would be plenty of space for Romney to promote himself as a Mr. Fixit capable of turning around the sluggish economy.

"This low tide gives Romney the chance to redefine himself," Axelrod noted in late April.

The Obama campaign's mission was to prevent that. In early April, addressing the American Society of Newspaper Editors in Washington, Obama let loose a withering assault on the latest version of Representative Paul Ryan's budget—which resembled the end-Medicare-as-we-know-it budget passed a year earlier by the House Republicans. The president's speech was reminiscent of the ones he delivered in December in Osawatomie, Kansas, and a year earlier at George Washington University. Obama cited Ryan's plan for drastic spending cuts and whopping tax breaks for the well-to-do (above and beyond those handed out by President George W. Bush) as the central doctrine of Republicans. He condemned the likely results: hundreds of thousands of children bounced out of Head Start, millions of mothers and kids cut from food nutrition programs, a lack of enforcement of food safety and clean air and water standards, a dramatic cutback in financial aid for college students, less medical research, reduced funding for veterans, a decline in Medicaid funding that would cause 19 million Americans to lose access to health care.

The House GOP budget, Obama ominously stated, "is a Trojan horse disguised as deficit reduction plans. It is really an attempt to impose a radical vision on our country. It is thinly veiled Social Darwinism." He noted that it would gut "the very things we need to grow an economy that's built to last: education and training, research and development, our infrastructure. It is a prescription for decline." And the president pointed out that Romney was aboard this train and had hailed the plan as "marvelous."

"In the months ahead," Obama said, "I will be fighting as hard as I know how for this truer vision of what the United States is all about." The Associated Press reported that Obama was in "combative campaign form." The vision war was on.

The next day Romney responded in kind. Speaking to the same group, he observed, "President Obama and I have very different visions for America." He blamed Obama for the lousy economy, but also blasted him for breaking a promise to keep unemployment below 8 percent (a vow Obama never made), for "apologizing for America abroad" (a charge Politifact.com had months earlier branded a "pants on fire" lie), for not ever proposing a serious deficit plan (though Obama had indeed offered a deficit reduction blueprint), for adding "nearly as much public debt as all the prior presidents combined" (an untrue statement), and for cutting $500 million from Medicare (which Politifact.com judged a "false" assertion). But Romney did agree the election "will be about principle."

ELECTIONS, THOUGH, RARELY REMAIN FOCUSED ON GRAND ABstract ideas. In the coming weeks, the campaign was dominated by a disappointing jobs report (in early May the unemployment rate dipped to 8.1 percent, but the economy added merely 115,000 jobs), the so-called Republican war on women (with the Romney camp disingenuously claiming that more women than men had lost jobs on Obama's watch and opportunistically leaping on Democratic strategist Hilary Rosen's inartful remark that Ann Romney, a mother of five, had "never worked a day in her life"), aging rocker (and Romney supporter) Ted Nugent's violent remarks about Obama, and scandals involving lavish spending at the General Services Administration and Secret Service agents consorting with prostitutes in Colombia.

Obama did try to push policy initiatives that illuminated his larger themes. He toured college campuses, touting his effort to continue low interest rates for student loans. (At one stop, Obama derided Republican Representative Todd Akin, who when asked if he would support legislation to prevent student loan interest rates from rising, had said, "America has got the equivalent of stage-three cancer of socialism because the federal government is tampering in all kinds of stuff it has no business tampering in.") The president called for a vote on the Buffett rule—which would require the wealthy to pay at least 30 percent of their income in taxes—but Senate Republicans mounted a successful filibuster against it.

Obama was running into opposition not only from obstructionist congressional Republicans and the Romney camp; so-called Super PACs and dark-money political outfits (empowered in part by the *Citizens United* decision) were starting to flood the presidential race with negative ads. The Karl Rove–backed American Crossroads released a video titled "Cool," which showed Obama dancing on *Ellen*, singing an Al Green song, and slow-jamming the news with Jimmy Fallon—an attempt to make the president appear superficial, though a viewer could be forgiven for sensing racial undertones in the ad. And Americans for Prosperity—a conservative outfit connected to the billionaire rightwing Koch brothers—dumped millions of dollars of anti-Obama ads into swing states. These and other conservative groups could be counted on to spend hundreds of millions of dollars to bash Obama in the months ahead.

THE OBAMA CAMP DID DESIRE A CLASH OVER VALUES, BUT IT had another goal as well: rip Romney apart. "We don't want to just counterpunch," Axelrod said.

In late April, the campaign released a Bill Clinton–narrated ad that praised Obama's decision to launch the bin Laden mission—and questioned whether Romney would have made the same tough call. "Shame on Barack Obama," Senator John McCain griped, for turning the killing of bin Laden "into a cheap political attack ad." But the ad was partly a response to Romney's accusation that Obama was weak and appeasing, and it was an indication that the president was willing to fight fiercely. (Obama craftily mounted a surprise trip to Afghanistan on the one-year anniversary of the OBL raid, applying the glow of that triumph to the much less popular war he was slowly winding down.)

It seemed a bit anticlimactic, but on the first weekend in May, Obama officially launched his reelection bid with rallies in the swing states of Virginia and Ohio. Speaking at Virginia Commonwealth University, before several thousand supporters, Obama attempted to reframe the election. The question, as he put it, was not whether voters were better off than they were four years ago, but if they would be better off in the coming years. He emphasized the economic fairness message he had developed over the course of the previous year.

But the red meat was Romney-bashing. Romney claimed that Obama had been disastrous for the economy. Obama countered that Romney and the Republicans *would be* disastrous should they gain power. Obama noted that congressional Republicans—who fancied "bigger tax cuts for the wealthiest Americans" and "deeper cuts to things like education and Medicare and research and technology"—had "found a champion" in Romney, who would "rubber-stamp this agenda." He portrayed Romney as an out-of-touch corporate overlord. Romney, he contended, "sincerely believes that if CEOs and wealthy investors like him make money, the rest of us will automatically prosper as well. . . . He doesn't seem to understand that maximizing profits by whatever means necessary—whether through layoffs or outsourcing or tax avoidance or union-busting—might not always be good for the average American or for our economy." Playing off a previous Romney statement, he added, "Corporations aren't people. People are people."

Some Democrats wondered if the president was mixing it up with Romney too soon. Incumbent presidents in the past often ignored the competition until the parties' nominating conventions in the summer. But Chicago was taking no chances. The polls showed a close race. Voters tended to like Obama more than Romney and considered the president more empathetic to the needs of middle-class Americans, but many believed Romney could handle the economy better. For Messina, Axelrod, and other top Obama advisers, it was not too early to start undermining Romney's most prominent strength.

But all of a sudden there was a detour. Appearing on *Meet the Press*, Vice President Joe Biden expressed his support for gay marriage. This off-the-cuff remark forced the president to do what he had already privately resolved to do: publicly reverse his opposition to same-sex marriage. He had previously told aides he intended to announce his change of heart, and the general sense in the White House was that this would occur sometime during the slow days of summer, prior to the Democratic convention. But Biden's remarks triggered a burst of questions and speculation, and Obama decided not to drag out his evolution. In a taped interview with a *Good Morning America* correspondent, he declared his support for gay marriage. ABC News broke into its daytime programming to report the news.

The episode—which happened the week following the kickoff of the reelection campaign—was a reminder that carefully planned messaging can be upended in a nanosecond by an unforeseen sound bite or development. Romney's camp learned this the hard way, too. As Romney attempted to pound Obama on the economy, he had to deal with a variety of not-in-the-script events. He had failed to challenge a supporter who called Obama a traitor—which made Romney seem like either a wimp or an enabler of rightwing kookiness. When the *Washington Post* published a lengthy and well-sourced account of his prep school days that included a troubling anecdote about the time he led a pack of students to assault a student who was thought to be gay, Romney implausibly claimed not to recall the incident. In the first fully Twitterized presidential election, neither campaign was able to control the daily narrative, and the early shadowboxing suggested that the race in the months ahead could quickly shift, perhaps due to tangential occurrences.

IN MID-MAY, THE OBAMA CAMPAIGN OPENED ITS FUSILLADE. It released a two-minute ad that would be broadcast in swing states, a six-minute video, and a website, each making the case that Romney's Bain Capital had engaged in predatory practices that left businesses it purchased in ruin, while Romney and his partners pocketed millions of dollars in fees. On the campaign trail, Romney was citing his business acumen as the primary reason he ought to be hired by American voters—claiming, without proof, that via Bain he had helped to create 100,000 jobs. (He talked little about his time as Massachusetts governor.) Yet this ad featured a steelworker from a plant Bain bought and then later shut down stating, "It was like a vampire. They came in and sucked the life out of us." (The day the campaign initiated its assault on Bain, Obama inexplicably held a fundraiser at the Blackstone Group, a hedge fund giant.)

Similar blame-Bain ads had worked in 1994 for Senator Ted Kennedy, who was nearly defeated by Romney. And anti-Bain assaults had slowed Romney's march to the nomination during the 2012 primaries, when Newt Gingrich lambasted Romney for having looted businesses and accused him of lying about creating 100,000 jobs while at Bain.

Messina and the gang were hopeful that the Bain ads would establish a leitmotif that would last until Election Day. When Romney ventured on a fundraising trip to Florida, the Obama campaign highlighted a medical instruments firm in Miami that after being bought by Bain shuttered two facilities, at a cost of 850 jobs, while Bain banked $242 million. Nearly four years after a Wall Street crash had triggered the recession, Obama and his aides wanted Romney to be regarded as a poster boy for the evil excesses of corporate America.

The Romney camp responded by charging that Obama was a foe of free enterprise. It was a continuation of a theme Romney had been pitching for a year: Obama didn't understand America and didn't believe it's an exceptional land. In other words, Obama is not truly an American at heart. It was a pale and more acceptable way of advancing the notion at the heart of birtherism (a discredited idea revived once more when Romney held a Las Vegas fundraiser with Donald Trump in May): Obama was the other. Romney and the Republicans just could not let go of this.

Several Democratic strategists wondered if the Bain attacks might backfire and cause voters to question Obama's dedication to free enterprise. On *Meet the Press,* Newark Mayor Cory Booker, an unconventional and often inspirational young Democratic leader, expressed disgust with negative attack ads, including Obama's Bain blasts. Though Booker quickly walked back his remarks, the Romney campaign immediately seized upon them—and Team Obama was placed on the defensive.

Yet Obama put the ensuing kerfuffle to good use. When he was asked about it at a press conference, he offered a better case than his surrogates had for why Romney's Bain connection was a legitimate and important campaign issue. After noting that the priority of a private equity firm is to maximize profits for investors—and that nothing was wrong with such an aim—the president declared, "that's not always going to be good for communities or businesses or workers." He continued:

> Governor Romney . . . is saying, "I'm a business guy and I know how to fix it," and this is his business. And when you're president, as opposed to the head of a private equity firm, then your job is not simply to maximize profits. Your job is to figure out how everybody in the country has a fair shot. . . . If your main argument for how to grow the economy is I knew how to make a lot of money for investors, then you're missing what this job is about. . . . My job is to take into account everybody, not just some. My job is to make sure that the country is growing not just now, but ten years from now and twenty years from now. This [Bain debate] is not a distraction. This is what this campaign is going to be about.

Obama was trying to outline a fundamental and personal distinction between himself and Romney. One candidate had a narrow what's-good-for-profits-is-good-for-all view; the other had an obligation to a more sweeping public interest.

At the same time, Romney kept declaring that Obama was a failure. As NBC News's *First Read* put it: "Obama's argument is that Romney doesn't have the values to be president (he'll look out only for the 1 percent, not the 99 percent); Romney's is that Obama doesn't have the skills (he's never run a business and his policies have failed)."

Obama headquarters opened up a related front in late May when it fired upon Romney's record as governor, noting that

Massachusetts had ranked forty-seventh nationwide in job creation when Romney was in charge. (But the unemployment rate
had been 4.7 percent, the Romney campaign replied.)

Senior Obama strategists, after examining focus group research, had concluded that the Massachusetts line of attack was
likely to be more effective than the Bain assault. It seemed that
portraying Romney as a vampire-like capitalist—even a mere four
years after Wall Street shenanigans triggered an economic free
fall—did not move independent voters as much as the argument
that Romney had been a lousy governor. The reelection campaign, which had aired the anti-Bain ads in five swing states, cut
Massachusetts-related spots for broadcast in at least ten states. Yet
subsequent polling did indicate the Bain ads could be an effective
attack on Romney, though pundits and other politerati continued
to debate this point. And the Obama crew, in response to new
media revelations about Romney's business dealings, intensified
the assault on Romney's Bain days and his secretive personal finances.

Meanwhile, Romney's case was bolstered in early June with a
dismal jobs report. Only 69,000 jobs had been created the previous month—less than half what most economists expected—and
unemployment ticked up a tenth of a point to 8.2 percent. This
was grim news for Obama (and the nation). For the third year in
a row, it looked as if the economy was heading toward a summer
dip, once again partly caused by European economic woes far
beyond Obama's control. In a press conference during which
Obama displayed his professorial side, going into details about
the impact of Europeans' financial troubles on the US economy,
he did revive his call for Congress to pass what remained of the
jobs bill he had proposed the previous September. But one off-
message statement—the "private sector is doing fine"—received
most of the media attention, and it was an open question whether

assailing Congress regarding the jobs bill leftovers would help him in his face-off against Romney.

The Romney campaign was banking on voter disappointment (or anger) with Obama and the economy. The Romney campaign wanted voters to use their vote in November to vent. His strategists certainly wouldn't mind an expression of ire. Obama was looking for another kind of voting. He wanted voters to see the election as a rational choice between two opposing sets of policies and values and then choose that set (and the fellow associated with it) that would better guide the nation in the years ahead. Not only were the two campaigns offering fundamentally opposing visions; they were each hoping for a different sort of election.

POLITICS IS AN ENTERPRISE OF MECHANICS. AS THE SUMMER OF 2012 began, Messina, Axelrod, and others in Chicago were crafting hyper-localized strategies for individual swing states. The reelection campaign had about 700 full-time employees—with half of them working for the digital department. "We're going to micro-target voters like never before and generate our own opportunities," Axelrod said, noting the campaign was developing new means to zero in on what he called "not high-information voters," meaning voters who hadn't paid a lot of attention to politics.

The Obama operation was also targeting Latino voters with more passion and sophistication than the Romney campaign. (Obama's decision to issue a directive preventing the deportation of young unauthorized immigrants boxed in Romney, who was on record supporting the deportation of all undocumented immigrants.) The campaign's volunteer army across the nation was at this point more extensive than Romney's (though the

Democratic ground game in Wisconsin had not won the day in the hotly contested gubernatorial recall election in early June). As Messina told reporters, he was building the "biggest grassroots effort in American political history." And the campaign's fundraising machine was whirring—but with Romney and his allies hauling in their own millions for what would be (thanks to *Citizens United*) the most cash-awash presidential race ever. It seemed a good bet that this would be the first election in US history in which an incumbent president would be outspent.

Politics is also an endeavor of impressions. Obama had spent much of 2011 establishing a narrative that distinguished his leadership and values from those of the Republicans. And the Republicans had returned the favor by tapping a quarter-billionaire who happily embraced the Ryan budget and Republican ubermessage. Ever since the Democrats' "shellacking" in November 2010, Obama had governed and politicked in accordance with this grand plan. Through one episode after the next—the tax cut tussle of the lame-duck session, the confrontations over the budget and the debt ceiling, the Libyan crisis—he had repeatedly demonstrated a guiding belief in strategic patience.

Time, however, was not an unlimited resource. The president had only several months left to convince voters that he deserved four more years. And he had to do so in a period of political instability. His health care overhaul was nearly eviscerated by the Supreme Court, but Republican-appointed Chief Justice John Roberts prevented what would have been a setback for the Obama campaign when he sided with the four liberals on the court to uphold the law. John Boehner had again declared there would be no future hike in the debt ceiling unless matched by an equal amount of budget cuts, threatening a replay of the previous summer's debt ceiling nightmare. (With the Bush tax cuts due to expire at the end of 2012, the national debt expected to reach the latest limit about then, and the automatic budget cuts set to hit in early 2013, the mother of all fiscal battles—pundits called it Taxmaggeddon—loomed on the horizon.) And the US economy appeared to be in an increasingly perilous position.

As the summer slog was beginning, Bill Clinton, the only Democrat since Franklin Delano Roosevelt to win two presidential elections, offered public advice to Obama: don't rely on a strategy centered on discrediting Romney. (He went off-message and remarked that Romney "had a sterling business career.") Instead, Clinton noted, with the election still easily cast as a referendum on Obama's record, the president had to transform the contest into a battle of ideas with Romney.

Obama was trying to shape the race that way. His foremost challenge now was to rise above the never-ending hurly-burly of modern politics and define starkly the central choice the country faced. His multiple confrontations with Boehner and the Tea Party–wagged Republicans had helped him set up the framework he desired: progressive values that emphasized communal principles and forward-looking investments pitted against conservative notions of unbridled markets and severe government downsizing. Still, the gravitational pull of the contest was toward an up-or-down referendum on the economy. Could Obama escape that force?

To do so, the forty-fourth president of the United States had to write a more nuanced and sophisticated political story for the nation and sell it to those in-between voters who held no strong loyalties to either side. That could end up being much more difficult that it had been to sell hope and change.

Acknowledgments

There's an old cliché about writing being a solitary endeavor, but no book is a solo project—this one especially. It was only possible to chronicle an ongoing and fast-changing story with the assistance of many others. At the top of the list is Siddhartha Mahanta, the chief researcher for this book. He compiled files on events as they were transpiring and diligently managed an ever-rising flood of research material. He was essential to the completion of this project. Matt Corley produced stunningly detailed chronologies of several episodes covered by this book. One former National Security Council official, upon seeing one of these time lines, said, "I wish I had someone like that working for me when I was at the White House." Asawin Suebsaeng provided valuable and meticulous fact-checking services and much-appreciated research support.

My agent Gail Ross is the godmother of this book. For years, she has always pushed me toward the next project and helped shape whatever that endeavor may be. This book springs from our ongoing collaboration—which has been enhanced by the participation of her colleague Howard Yoon.

Henry Ferris edited this book with passion and commitment. At our first meeting, he understood the project long before I finished the pitch, and as we rushed to finish, he sacrificed holiday time and toiled hard to guarantee that the book would be as good as it could be. Laurie McGee, applying her copyediting skills, fine-tuned the manuscript. I must also thank the team at William Morrow for all the support: Liate Stehlik, Lynn Grady, Seale Ballenger, and Andy Dodds.

I am fortunate to have a wonderful day (and night) job managing the Washington bureau of *Mother Jones*. I am indebted to editors Monika Bauerlein and Clara Jeffery for allowing me to escape some of my daily duties to work on this book. I thank Madeleine Buckingham and Steven Katz, respectively the chief executive officer and publisher of *Mother Jones,* for all their support before, during, and after this project. I particularly want to express appreciation for my colleagues in the Washington bureau: Daniel Schulman, Nick Baumann, Stephanie Mencimer, Kate Sheppard, James Ridgeway, Tim Murphy, Andy Kroll, and Adam Serwer. They provide camaraderie and produce impressive journalism every day. It is an honor and a pleasure to work with these talented reporters and editors. (Schulman gets extra credit for reading the manuscript and recommending revisions.) Also, many thanks to all my other colleagues at the magazine and to Phil Straus, Adam Hochschild, and other members of the board of directors.

For the past few years, I've been privileged to be part of the MSNBC community—and that was a great benefit while I was writing this book. The green room is a valuable spot for reporting, and jousting on-air about many of the matters included in this work helped me develop and refine my own thinking about these events. I have many people at the network to thank: Phil Griffin, Rachel Maddow, Lawrence O'Donnell, Ed Schultz, Chris Jansing, Martin Bashir, Al Sharpton, Izzy Povich, Pamela Stevens, Amy Shuster, Shannah Goldner, Natasha Lebedeva, Gregg Cockrell, and all the other hosts, producers, and bookers. A very special tip of the hat goes to the gang at *Hardball,* which often has seemed like a home away from home: Chris Matthews, John Reiss, Ann Klenk, Tina Urbanski, Connie Patsalos, Querry

Robinson, Colleen King, Erin Delmore, Michael LaRosa, Robert Zeliger, Derbin Cabell Jr., Chester Reis, Gary Lynn, Maria Sevilla, Rose Procopio, Carla Dakin, Aleese Majeed, George Tolman, Carl Trost, and everyone else. Also, special thanks to Howard Fineman and Michael Steele for being particularly good on-air partners.

I am grateful for the ongoing comradeship and encouragement (online and off) supplied by several journalist pals: Lynn Sweet, Sam Stein, Bill Press, George Condon, Susan Page, Jack Shafer, Roger Simon, James Pinkerton, Arianna Huffington, Robert Wright, and, most notably, Michael Isikoff, my onetime coauthor.

Through the ten months it took to write this book, I enjoyed the support of family members and friends, even though I was often not in the position to reciprocate fully. I thank them all: Ruth Corn Roth, Gordon Roth, Steven Corn, Amy Corn, Barry Corn, Reid Cramer, Sonya Cohen, Sally Kern, Stephen Kern, Marco DiPaul, Jenny Apostol, Sam Kittner, Bobbi Kittner, Andrew Steele, Katja Toporski, Louis Spitzer, Gillian Caldwell, James Grady, Bonnie Goldstein, Ricki Seidman, Eric Scheye, David Williams, Robert Shapiro, Elizabeth Nessen, Micah Sifry, Marc Cooper, Joe Pichirallo, Beth Broderick, Bertis Downs, Mike Mills, Ellen Barkin, Joseph Finder, Mary Ann Akers, Stephanie Slewka, Horton Beebe-Center, Henry Von Eichel, Julie Burton, Steve Earle, Allison Moorer, Jill Sobule, Jamie Kitman, Julian Borger, Tony Alfieri, John Marttila, Harry Shearer, Tom Watson, Francie Randolph, Carlotta Luke, Christopher Luke, Laura Flanders, Elizabeth Streb, and Steven Prince.

A special shout-out, as always, goes to Peter Kornbluh.

Several people were quite helpful but do not want to be named. You know who you are.

Finally, and most of all, I owe much—gratitude being but a small slice of it—to my loving family: Welmoed, Maaike, and

Amarins. This book begins with you on the dedication page and concludes with this wholehearted thank-you, and that is appropriate. Everything starts and ends with the three of you. Welmoed was a complete partner in the endeavor, providing constant encouragement, as well as insightful critiques of the under-construction manuscript that greatly improved the book. Her contribution to this project—as with all else in our lives— was immense. I am profoundly grateful for the love and joy she has shared with me. While I was working on this book, Maaike and Amarins managed to become more mature, intelligent, de-lightful, and engaged in the world, somehow without their fa-ther's complete attention. Their steady interest in this project was always uplifting. The best part of ending this chapter for our family is being able to write the next pages together.

A Note on Sources

History is never finished. There is always more information to unearth. That's especially true with recent history, when
the dust hasn't settled and the lead players remain in their roles.

This book is the first telling of a tumultuous year in US politics and chronicles events that were under way as it was being
reported and written. *Showdown* is primarily based on more
than a hundred interviews with key participants: past and present White House officials, Democratic and Republican members of Congress, political strategists, Capitol Hill staffers from
both parties, and others. These sources often speak to reporters
and authors on background—meaning they cannot be quoted
by name. This is a common, though sometimes regrettable,
practice in journalism, but one with an obvious trade-off. It is
usually impossible to penetrate the bastions of power and influence in Washington (or elsewhere) without depending on un-
identified sources—especially when the topics at hand are recent
decisions and actions for which there are no official records to
examine.

Of course, it's best when sources are completely identified,
and whenever possible I have named sources. But without relying on confidential sources, the inside story of an ongoing
presidency could not be told. Many government officials are not
allowed to discuss publicly their experiences and observations.
In some cases, sources requested anonymity because speaking

freely about the president and the White House (or members of Congress) could harm their professional relationships and damage their careers.

When I recount private meetings and conversations, these accounts are generally based on interviews with people who witnessed these events or learned of them directly from a participant. In several instances, participants consulted notes they had taken during those meetings before discussing these events with me; at other times sources shared such notes with me. When I describe the thoughts or opinions of a particular official, these passages are based on interviews with one or more sources who possessed direct knowledge of that information—or on public comments that explicitly indicated what a person was thinking.

Though there are no books yet published that span the full period covered by this book, there are several works that I found highly useful for understanding Barack Obama and his presidency.

Foremost among those was Jonathan Alter's *The Promise: President Obama, Year One* (Simon and Schuster, 2010). Richard Wolffe's *Revival: The Struggle for Survival Inside the Obama White House* (Crown, 2010) offered a valuable deep dive into the White House during the first few months of 2010. Ron Suskind's controversial *Confidence Men: Wall Street, Washington, and the Education of a President* (HarperCollins, 2011) probed the crucial nexus of politics and economic policy.

In *The Audacity to Win: How Obama Won and How We Can Beat the Party of Limbaugh, Beck, and Palin* (Penguin Books, 2010), David Plouffe presented a gripping, from-the-inside account of the 2008 campaign that revealed much about the president and his top advisers. David Remnick's *The Bridge: The Life and Rise of Barack Obama* (Random House, 2010) was a penetrating and perceptive biography that tied Obama's past to the present. But no book may better explain the young Obama than his own elegant *Dreams from My Father: A Story of Race and Inheritance* (Crown, 2007).

As noted above, much of this book is drawn from interviews. But I also benefited from the fine journalism conducted by reporters who covered the events depicted in this book as they were occurring. Two daily online political newsletters were especially valuable: "First Read," compiled by the political unit of NBC News, and "Playbook," written by *Politico*'s Mike Allen. Each provides a blow-by-blow rundown of the prominent political and policy stories of Washington (and compiles the must-read news stories of the day). RealClearPolitics, a site that aggregates political articles (while offering original reporting and tracking polls), was also quite useful.

There are specific reporters to whom I owe thanks for either particular articles or ongoing coverage of a specific story. John Heilemann, the coauthor of *Game Change: Obama and the Clintons, McCain and Palin, and the Race of a Lifetime* (HarperCollins, 2010), the go-to account of the 2008 presidential campaign, wrote an excellent piece in *New York* magazine revealing and evaluating post-midterms shifts at the White House ("The West Wing, Season II," January 23, 2011).

No one covered the New START ratification battle better than Josh Rogin, who writes "The Cable" blog for *Foreign Policy*. He broke news and provided much insight on the complicated twists and turns, and explained the policy and technical intricacies of the ratification fight. The *New York Times*' Peter Baker wrote a fine ticktock covering this skirmish ("Obama Gamble Pays Off With Approval of Arms Pact," December 23, 2010).

Ryan Lizza's marvelous examination in *The New Yorker* of the impact of the Arab Spring on Obama's foreign policy—which disclosed Obama's early Presidential Study Directive concerning the possibility of democratic change in North Africa and the Middle East—was an incisive review of the president's decision making during the Egyptian and Libyan uprisings ("The Conse-quentialist," May 2, 2011).

Politico's David Rogers, a veteran journalist who has long re-ported on Washington's budget wars, covered the near govern-ment shutdown and subsequent debt-ceiling showdown with more knowledge (and better sources) than perhaps any other reporter. His work was essential for understanding the complexi-ties of each of those squabbles. Lori Montgomery of the *Washing-ton Post* also superbly covered those confrontations and produced sharp daily accounts (often working with her colleagues at the newspaper). One particularly valuable wrap-up of the debt-ceiling clash was reported and written by her, Paul Kane, Brady Dennis, Alec MacGillis, David Fahrenthold, Rosalind Helderman, Feli-cia Sonmez, and Dan Balz ("Origins of the Debt Showdown," August 6, 2011).

During the 2011 budget and debt-ceiling tussles, the Huff-ington Post mounted live blogs that published up-to-the-second reports. Afterward, these running accounts served as handy chronologies—as did a daily explainer that was produced during the debt-ceiling fracas by my colleagues at *Mother Jones* for our website. Talking Points Memo also provided valuable daily—sometimes hourly—coverage of the budget and debt-ceiling bat-tles and the subsequent fight over the payroll tax cut. Ditto for Greg Sargent's blog, "The Plum Line," at the *Washington Post*. Ezra Klein's "Wonkblog" column for that newspaper was an ever-useful source of analysis of budget policy.

In the days immediately following the Osama bin Laden raid, several news organizations quickly produced riveting accounts of the daring operation. Mark Mazzetti, Helene Cooper, and Peter Baker handled this assignment for the *New York Times* ("Behind the Hunt for Bin Laden," May 2, 2011). Adam Goldman and Matt Apuzzo did the same for the Associated Press ("Osama Bin Laden Dead: How One Phone Call Led U.S. to Bin Laden's

Doorstep," May 2, 2011). Greg Miller and Joby Warrick covered this territory for the *Washington Post* ("Bin Laden Discovered 'Hiding in Plain Sight,'" May 3, 2011). Marc Ambinder at the *National Journal* wrote one of the first behind-the-scenes articles ("The Secret Team That Killed bin Laden," May 2, 2011). Another good account of the raid and the intelligence work that led to the mission later appeared in *Target: Bin Laden: The Death and Life of Public Enemy Number One,* an e-book produced by ABC News.

All official remarks made by President Barack Obama can be found at www.whitehouse.gov.

INDEX